CERTIFIED
ISA
ARBORIST®

Arborists' Certification Study Guide

By Sharon J. Lilly

The *Arborists' Certification Study Guide* is intended to serve as a recommended program of study. It is not intended to be the only program of study to obtain the ISA Certified Arborist credential. The narrative portion of this *Guide* is general in nature, to serve as a primer. The practices and recommendations contained in this *Guide* should be used in practice only after a thorough analysis of the particular situation. The items discussed herein should be used or practiced only by those properly trained, educated, and experienced in the field of arboriculture. ISA is responsible only for the educational program contained in this *Guide* and not for the use or misuse of these ideas in specific field situations or by inexperienced or improperly trained individuals.

The ISA seal is a registered trademark.

Editorial and Production Manager: Peggy Currid
Composition and Cover Design: Amy Reiss
Printed by: ADR BookPrint, Wichita, KS

International Society of Arboriculture
P.O. Box 3129
Champaign, IL 61826-3129
Web site: www.isa-arbor.com
E-mail: isa@isa-arbor.com

10 9 8 7
ISBN 1-881956-26-1
01-03/CO/2000

ACKNOWLEDGMENTS

The author wishes to thank the following individuals for contributing or suggesting text passages to be included in this *Guide*:

Dr. Peter Donzelli
Randy Miller
Dr. Jim Clark
Dr. Kim Coder
Dr. Donald Ham
Denice Britton
Don Blair

Illustrators

Bryan Kotwica
Mike Thomas
Todd Möell
Todd Akers
Glenn Wasserman
Les Powers

The following are gratefully acknowledged for contributing artwork and photographs:

Bonnie Appleton
ArborCare, Inc.
John Ball
Bartlett Tree Research Laboratories
Jim Clark
ChemLawn Services
Larry Costello
Michael Dirr
Ed Gilman
Ed Irish
Warren Johnson
Bobby Joyner
Jerry King
Gary Kling
Bill Kruidenier
Lauren Lanphear
Sharon Lilly
Nelda Matheny
National Arbor Day Foundation
National Arborist Association
OSU Curriculum Materials Service
Bob Partyka
Chuck Powell
Dave Shetlar
Alex Shigo
Tom Smiley
T. Davis Sydnor
Vermeer Manufacturing
Gary Watson

The International Society of Arboriculture and the author of this *Guide* wish to thank and acknowledge the following professionals for taking time to review parts or all of the *Guide*:

Mark Adams
Dr. Douglas Airhart
Dr. Bonnie Appleton
Jennifer Arkett
Dr. John Ball
Dr. Nina Bassuk
Don Blair
Joe Bones
Ed Brennan
Dr. Beth Buchanan
Dr. Bill Chaney
Mark Chisholm
Dr. Kim Coder
Dr. Larry Costello
Dr. Phillip Craul
Dr. Peter Donzelli
Dr. Bruce Fraedrich
Dr. Roger Funk
Tim Gamma
Peter Gerstenberger
Frank Gifford
Dr. Ed Gilman
Lisa Ward Grant
Bruce Hagen
Dr. Don Ham
Dr. Mark Harrell
Jim Ingram
Jeff Jepson
Dick Jones
Dr. Bobby Joyner
Lauren Lanphear
Dave Leonard
John Lloyd
Nelda Matheny
Dr. Fredric Miller
Randall Miller
Dr. Robert Miller
Dave Mooter
Dwayne Neustaeter
Dr. David Nielsen
Ken Palmer
Jim Patterson
Ward Petersen
William Rutherford
Dr. Dennis Ryan III
Jim Skiera
Dr. Tom Smiley
Rip Tompkins
Trevor Vidic
Dr. Gary Watson

The author also wishes to thank Peggy Currid, Kathy Ashmore, Phyllis Picklesimer, Amy Reiss, and Mary Pelletier-Hunyadi for their assistance in the production of the *Guide*.

OTHER RECOMMENDED RESOURCES

*American National Standards Institute. ***American National Standard for Tree Care Operations—Tree, Shrub and Other Woody Plant Maintenance—Standard Practices*** (A300). ANSI, New York, NY.

*American National Standards Institute. ***American National Standard for Tree Care Operations—Tree, Shrub, and Other Woody Plant Maintenance—Standard Practices (Fertilization)*** (A300, Part 2). ANSI, New York, NY.

*American National Standards Institute. ***American National Standard for Tree Care Operations—Tree, Shrub, and Other Woody Plant Maintenance—Standard Practices (Support Systems a. Cabling, Bracing, and Guying*** (A300, Part 3). National Arborist Association, Manchester, NH.

*American National Standards Institute. ***American National Standard for Tree Care Operations—Pruning, Trimming, Repairing, Maintaining, and Removing Trees and Cutting Brush—Safety Requirements*** (Z133.1). International Society of Arboriculture, Champaign, IL.

American National Standards Institute. ***American National Standard for Nursery Stock*** (Z60). American Association of Nurserymen, Washington, DC.

*Costello, L.R. 2000. ***Training Young Trees for Structure and Form*** (videocassette and booklet). University of California Cooperative Extension, Davis, CA.

Craul, P.J. 1999. ***Urban Soils: Applications and Practices.*** Wiley and Sons, New York, NY.

*Dirr, M.A. 1998. ***Manual of Woody Landscape Plants.*** 5th ed. Stipes Publishing, Champaign, IL.

*Gilman, E.F. 1997. ***An Illustrated Guide to Pruning Trees***. Delmar Publishers, Albany, NY.

*Gilman, E.F. 1997. ***Trees for Urban and Suburban Landscapes***. Delmar Publishers, Albany, NY.

*Gilman, E.F. 1998. ***Horticopia: Trees, Shrubs and Groundcovers*** (CD-ROM). 2nd ed. Horticopia, Purcellville, VA

*Harris, R.W., J.R. Clark, and N.P. Matheny. 1999. ***Arboriculture: Integrated Management of Landscape Trees, Shrubs, and Vines***. 3rd ed. Prentice Hall, Upper Saddle River, NJ.

*Horticopia, Inc. 2001. ***Horticopia Professional (Arborist Edition)*** (CD-ROM). Horticopia, Purcellville, VA.

*International Society of Arboriculture. 1995. ***Tree-Pruning Guidelines***. ISA, Champaign, IL.

*International Society of Arboriculture. ***ArborMaster Training Video Series I: Climbing Techniques and Equipment*** (six videocassettes and workbooks). ISA, Champaign, IL.

*International Society of Arboriculture. ***ArborMaster Training Video Series II: Climbing Innovations*** (two videocassettes and workbooks). ISA, Champaign, IL.

*International Society of Arboriculture. ***ArborMaster Training Video Series III: Chain Saw Safety, Maintenance, and Cutting Techniques*** (six videocassettes and workbooks). ISA, Champaign, IL.

*International Society of Arboriculture. ***ArborMaster Training Video Series IV: Rigging*** (seven videocassettes and workbook). ISA, Champaign, IL.

*International Society of Arboriculture and National Arborist Association. 1999. ***Basic Training for Tree Climbers*** (five videocassettes and workbook). ISA, Champaign, IL, and NAA, Manchester, NH.

*Jepson, J. 2000. ***The Tree Climber's Companion.*** 2nd ed. Beaver Tree Publishing, Longville, MN.

*Johnson, W.T., and H.H. Lyon. 1991. ***Insects That Feed on Trees and Shrubs.*** 2nd ed. Comstock Publishing Associates, Cornell University Press, Ithaca, NY.

Kramer, P.J. 1969. ***Plant and Soil Water Relationships: A Modern Synthesis***. McGraw-Hill, New York, NY.

Kramer, P.J., and T.T. Kozlowski 1979. ***Physiology of Woody Plants***. Academic Press, New York, NY.

*Lilly, S. 1998. ***Tree Climbers' Guide***. International Society of Arboriculture, Champaign, IL.

*Lloyd, J., Editor. 1997. ***Plant Health Care for Woody Ornamentals: A Professional's Guide to Preventing and Managing Environmental Stresses and Pests***. Cooperative Extension Service, University of Illinois at Urbana-Champaign, Urbana, IL, and International Society of Arboriculture, Champaign, IL.

*Matheny, N.P., and J.R. Clark. 1994. ***A Photographic Guide to the Evaluation of Hazard Trees in Urban Areas.*** 2nd ed. International Society of Arboriculture, Champaign, IL

*Matheny, N.P., and J.R. Clark. 1998. ***Trees and Development***. International Society of Arboriculture, Champaign, IL.

*Merullo, V.D., and M.J. Valentine. 1992. ***Arboriculture and the Law***. International Society of Arboriculture, Champaign, IL.

*National Arborist Association and International Society of Arboriculture. 2000. ***Basic Training for Ground Operations in Tree Care*** (five videocassettes and workbook). NAA, Manchester, NH, and ISA, Champaign, IL.

*Shigo, A.L. 1986. ***A New Tree Biology***. Shigo and Trees, Associates, Durham, NH.

*Sinclair, W.A., H.H. Lyon, and W.T. Johnson 1987. ***Diseases of Trees and Shrubs***. Comstock Publishing Associates, Cornell University Press, Ithaca, NY.

Sunset Editors. 1995. ***Western Garden Guide.*** 6th ed. Sunset Publishing, Menlo Park, CA.

*Watson, G.W., and E.B. Himelick. 1997. ***Principles and Practice of Planting Trees and Shrubs***. International Society of Arboriculture, Champaign, IL.

*Watson, G.W. No date. ***Root Injury and Tree Health*** (videocassette and booklet). Illinois Arborist Association, Antioch, IL.

Whitcomb, C.E. 1991. ***Establishment and Maintenance of Landscape Plants***. Lacebark, Stillwater, OK.

*Items marked with an asterisk are available for purchase from ISA. Call 1-888-ISA-TREE, or order on-line at www.isa-arbor.com.

TABLE OF CONTENTS

Chapter 9: Tree Support and Protection Systems

Chapter 10: Diagnosis and Plant Disorders

Chapter 11: Plant Health Care

Chapter 12: Tree Assessment and Risk Management

Chapter 13: Trees and Construction

Chapter 14: Safety

Chapter 15: Climbing and Working in Trees

NOTICE TO EXAM CANDIDATES

The purpose of the ISA Certified Arborist examination is to assess a candidate's knowledge and skills in the field of arboriculture. The examination is designed to assess the fundamental knowledge and skills that all tree care professionals should have, regardless of their area of practice. Some knowledge is gained cumulatively through time and experience in the field and through the accumulation of scientific information related to the care of trees. Other things can be learned through the study of texts or in the classroom. The breadth of knowledge required to pass the exam is great; thus, candidates should begin preparing for the exam well in advance of the scheduled test administration date.

The Certified Arborist examination comprises, and tests candidate knowledge in, the following twelve areas of arboriculture: (1) Biology; (2) Identification and Selection; (3) Soil and Water; (4) Risk Assessment; (5) Tree Nutrition and Fertilization; (6) Plant Health Care; (7) Installation and Establishment; (8) Pruning; (9) Cabling, Bracing and Lightning Protection; (10) Problem Diagnosis; (11) Construction and Preservation; and (12) Safe Work Practices. It is important to note that different sets or groups of test questions are used each time the certification examination is administered, so the content of the ISA examination is never exactly the same in any two administrations.

The ISA *Arborists' Certification Study Guide* is designed to help ISA Certified Arborist candidates review the topics covered on the examination. The *Study Guide* provides useful information on examination content, practice items, and preparation for testing. *All Certified Arborist candidates should also be aware, however, that the* Guide *does not necessarily represent the full range of examination content, nor does it necessarily present the full range of individual question difficulty included on any actual examination.*

The *Study Guide* also includes a recommended resource list that is intended to further assist candidates in developing a knowledge base useful to a professional arborist. *However, all certification candidates should be aware that the reference list does not attempt to include all acceptable reference material, nor is it suggested that the ISA certification examination is necessarily based on these references.*

The ISA Certification Board of Directors and certification program do not endorse, support, or assist in the development of materials other than this Guide, *including those materials listed in the reference section of the* Guide. *In addition, ISA does not and cannot guarantee enhanced performance on the certification examination as a result of using this* Guide.

INTRODUCTION

Each chapter of the *Study Guide* consists of five sections: narrative, workbook questions, challenge questions, sample test questions, and other sources of information. In addition, each chapter includes a list of objectives and a list of the key terms introduced.

Before beginning to read the narrative section, the reader should review the objectives and look over the list of key terms. Doing so provides a focus and orients the reader toward the concepts that should be learned in that chapter. The next step is to read through the narrative section. The purpose of this section is to serve as a primer and provide the basic concepts of the topic. Many illustrations, photographs, and charts are included to help the reader understand the material.

The workbook section follows the narrative. The questions in this section are designed to reinforce the basic concepts presented. These short-answer questions help the reader determine which parts of the chapter require more study.

Once the workbook section has been completed successfully, the reader should proceed to the section containing other sources of information. These readings provide greater depth and detail on each subject and help the reader apply the basic concepts to field situations.

The goal of the challenge questions is to strengthen the reader's understanding of the subject areas and to apply the concepts learned. These questions take the reader beyond the scope of the narrative. To answer them, the reader must demonstrate a more detailed understanding of the subject and must be able to project the concepts into practical applications.

The final section is the sample test questions. These questions are typical of those that might appear on the certification exam. Their purpose is to help familiarize the reader with the style and scope of the exam. The questions on the certification exam are based on, but not limited to, the material presented in the *Study Guide*. Some exam questions are based on the other sources of information provided. Other questions reflect knowledge that an arborist would be expected to gain from experience. Many of the questions test the reader's ability to apply the concepts learned in the *Study Guide*.

CHAPTER 1

TREE BIOLOGY

CHAPTER 1 TREE BIOLOGY

CHAPTER 1
TREE BIOLOGY

Objectives

1. Learn the structures and functions of the buds, leaves, wood, and roots of a tree.
2. Understand the interaction of structure and function in tree biology.
3. Learn the basic composition of a tree's vascular system and understand how water and carbohydrates are transported within this system.
4. Describe the relationship roots have with mycorrhizae. Explain how the soil environment affects root growth and distribution.
5. Understand the processes of photosynthesis and respiration and the factors affecting these processes.
6. Understand how the growth and development of a tree is the result of the interaction between its genetic potential and the environmental surroundings.
7. Explain the concept of Compartmentalization Of Decay In Trees (CODIT).

Key Terms

abscission zone
absorbing roots
adventitious buds
anthocyanins
antitranspirant
apical bud
apical dominance
apical meristem
auxin
axial transport
axillary bud
branch bark ridge
branch collar
buds
cambium
carbohydrate
carotenoids
chlorophyll
chloroplast
CODIT
companion cell
compartmentalization
cork cambium
cuticle
cytokinin
deciduous
decurrent
differentiation
diffuse porous
dormant
epicormic
evergreen
excurrent
fiber
geotropism
growth rings
guard cells
gymnosperm
heartwood
included bark
internode
lateral roots
lenticel
meristem
mycorrhizae
node
osmosis
parenchyma cells
petiole
phloem
photosynthate
photosynthesis
phototropism
radial transport
ray
reaction zone
respiration
ring porous
sapwood
shakes
sieve cells
sieve tube elements
sink
sinker roots
source
stomata
symbiosis
tap root
terminal bud
tracheid
transpiration
tropism
vessels
xylem

INTRODUCTION

Trees, by most definitions, are woody plants that are large and have a single main trunk. To paraphrase Alex Shigo, trees are long-lived perennials that are woody and compartmentalizing organisms. This series of unique characteristics has allowed trees to dominate the vegetation of large areas of the world.

Arboriculture is both an art and a science. It combines skill and craft with knowledge and fact. Thus, a foundation for the practice of arboriculture is a thorough understanding of how trees grow.

The skilled arborist learns how a tree grows and how to care for and manage it in a way that supports its growth and development. Like physicians, arborists use knowledge of tree growth and development to diagnose problems, assess potential, and prescribe treatments. And, just as in medicine, prescription before diagnosis is malpractice; the arborist must understand tree biology before embarking on a program of care.

The study of tree biology is the study of structure and function and the relationship between them. Anatomy and morphology are the studies of the component parts of the tree. Physiology is the study of the biological and chemical processes within these structures, providing the basis for function. This text will focus primarily on the biology of hardwood trees, but some attention will be given to conifers and palms.

TREE ANATOMY

Basic Structure: Cells and Tissues

All living organisms share a basic organizational theme, based upon cells, tissues, and organs. Cells are the basic building blocks of structure. New cells arise from the division of existing cells. In trees, this process occurs in specialized zones called **meristems**. Following division, cells undergo **differentiation**, which changes their structure and permits them to assume a variety of specific functions. Cells with similar structure and function are arranged into tissues such as bark or wood. Tissues are then organized into organs, of which plants have five: leaves, stems, roots, flowers, and fruit. Finally, organs are organized into intact, fully functional organisms—trees.

There are two basic types of meristems: 1. primary, which produce the cells that result in elongation of shoots and roots, and 2. secondary (also known as lateral meristems), which produce cells that result in increases in diameter. The presence of secondary meristems, growing within the stems and branches and producing wood, allows trees to grow so large. The exceptions to this pattern are palms, which lack secondary growth.

Meristems located at the ends of shoots and roots are primary, or **apical, meristems**. In shoots, they are found inside **buds** (Figure 1.1). The overlapping scales or modified leaves of buds protect both the meristematic region and the developing shoot. In roots, the meristem is protected by a root cap.

Trees have two secondary or lateral meristems. The first lateral meristem is the **cambium**. The cambium is a thin sheath of dividing cells that produces the cells that will become the vascular system of the tree. The cambium produces two kinds of tissue: **xylem** to the inside, and **phloem** to the outside (Figure 1.2). The second lateral meristem is the **cork cambium**, which produces the bark (Figure 1.3).

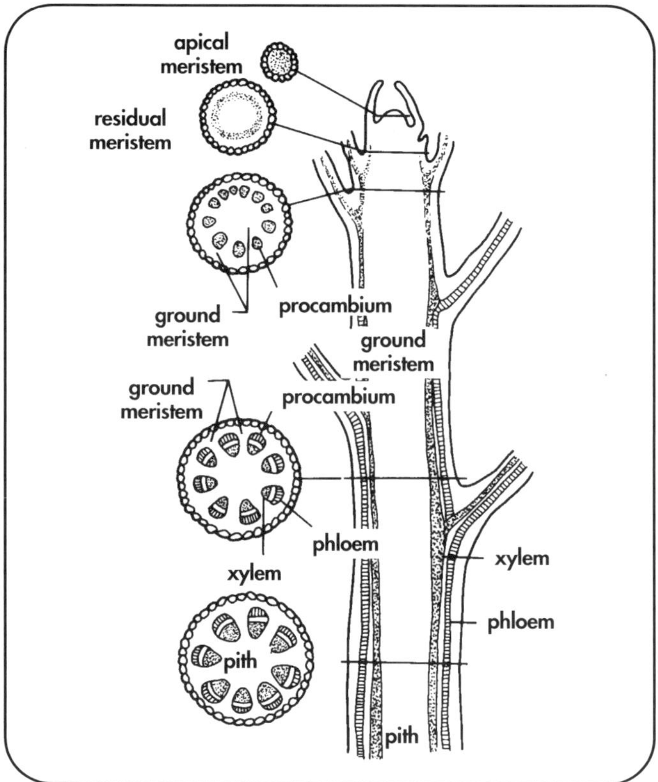

Figure 1.1 Longitudinal and cross sections through a shoot tip.

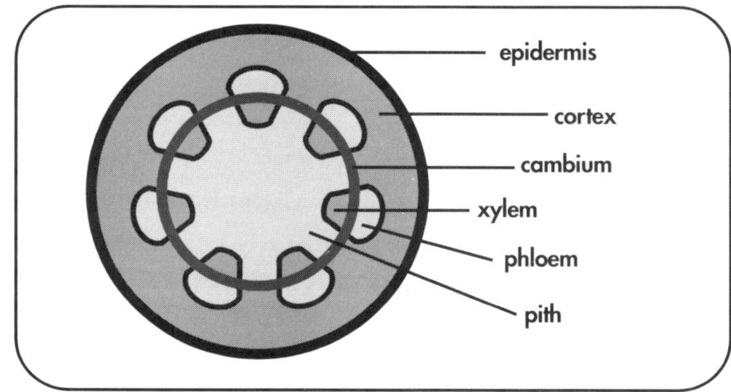

Figure 1.2 Cross section of a young stem.

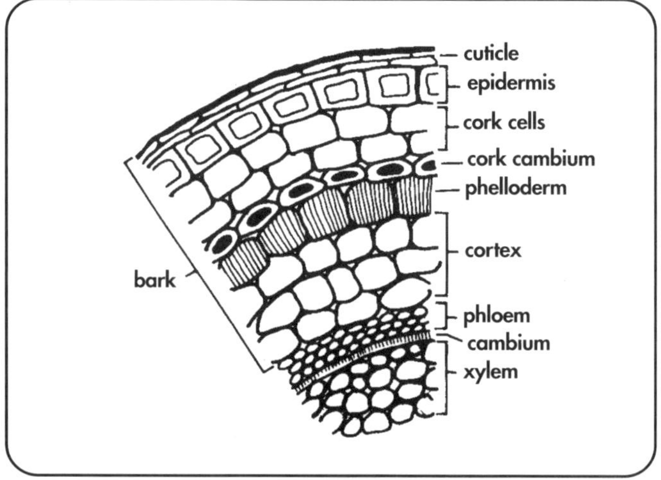

Figure 1.3 Partial cross section through the bark, phloem, and cambium, and slightly into the wood of a hardwood tree.

Xylem and Phloem

The xylem, or wood of the tree, has four primary functions: 1. conduction of water and dissolved minerals (elements), 2. support of the weight of the tree, 3. storage of carbohydrate reserves, and 4. defense against the spread of disease and decay.

Xylem is a complex tissue, composed of both dead and living cells. Xylem of **gymnosperms** (for example, pines and spruces) is composed of **tracheids**, **fibers**, and **parenchyma cells**. Tracheids are elongated, dead cells with pointy ends and thickened walls, which function in water conduction and mechanical support. Fibers provide mechanical strength. Parenchyma cells are living cells interspersed among the other xylem cells. Parenchyma cells located in the outer layers of xylem are active in water conduction. Parenchyma cells also store carbohydrates and help defend against decay.

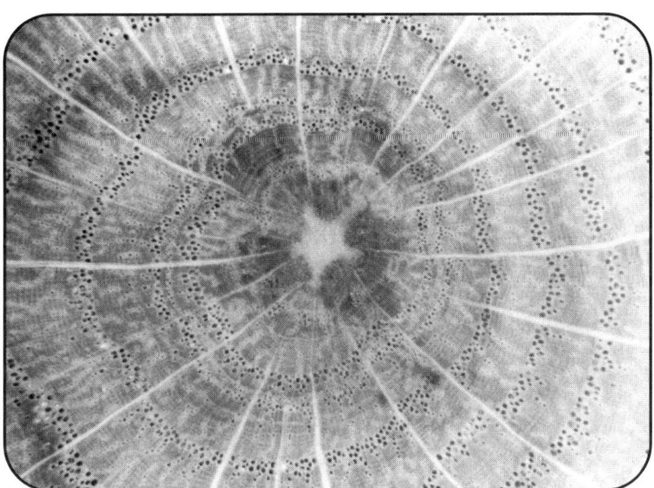

Figure 1.4 Cross section through a California black oak *(Quercus kelloggii)*, a ring-porous tree. Vessels formed in the spring are large; vessels formed later are small. Note the prominent rays and star-shaped pith, characteristic of oak.

The xylem of hardwood trees is also made up of tracheids, fibers, and parenchyma cells, as well as **vessels**. Vessels are the primary conducting elements in hardwoods. The vessels can be thought of as stacks of dead, hollow cells that form long tubes of water-conducting elements. Vessels are much more efficient in water conduction than are tracheids. Another differentiating factor is that parenchyma cells are more abundant in hardwood trees and are arranged close to the vessel elements.

The physical and biological properties of different types of trees are related to the arrangement of the cell types within the xylem. Some trees form wide vessels early in the growing season and narrower vessels later in the season. These trees are said to be **ring porous** (Figure 1.4), and they include elms (*Ulmus*), oaks (*Quercus*), and ashes (*Fraxinus*), among others. Other species produce vessels of uniform size throughout the growing season and are called **diffuse porous** (Figure 1.5). Examples include maples (*Acer*), planetrees (*Platanus*), poplars (*Populus*), and others. When a tree has been cut and can be viewed in cross section, **growth rings** (Figure 1.6) are visible in the xylem. These rings are the result of seasonal production of xylem by the cambium. They appear as rings because the relative size and density of the vascular tissues changes throughout the growing season. As the season progresses, cells become smaller in diameter. Thus, the contrast between cells produced early in the season (earlywood) and those produced later (latewood) allows the diameter increase within an individual year to be seen.

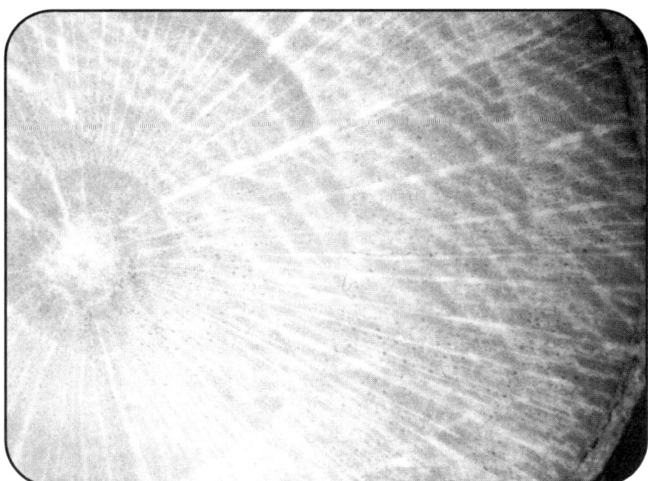

Figure 1.5 Cross section through American beech *(Fagus grandifolia)*, a diffuse-porous species. The vessels are small and scattered evenly throughout each yearly increment of growth.

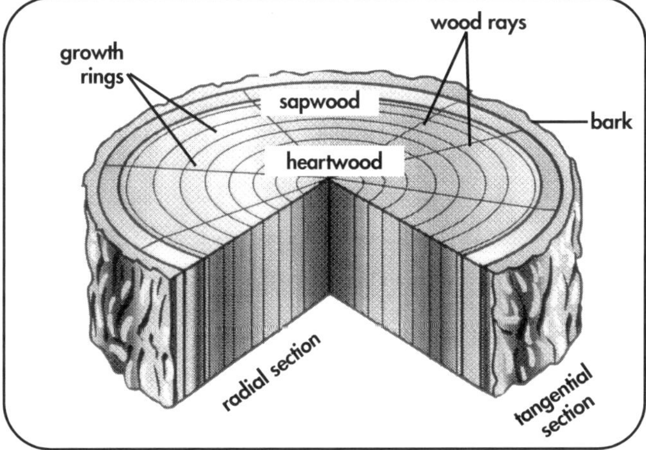

Figure 1.6 Cross section of wood showing growth rings, sapwood, and heartwood.

Not all of the conducting elements in the xylem transport water. In conifers, two to twelve rings may actually conduct water. In trees such as elm, only the outermost one or two rings do so. Xylem that conducts water is called **sapwood**. There are many living parenchyma cells in the sapwood. Farther inside the tree is the **heartwood**, nonconducting tissue that is sometimes darker in color than the sapwood.

The phloem is responsible for the movement of sugars, produced in the leaves, to other plant parts. The phloem carries sugar to the roots and throughout the plant for storage or consumption. Movement of sugar in the phloem is relatively slow and occurs along pressure gradients. Phloem transport requires energy. Unlike the hollow vessel elements and tracheids of the xylem, the phloem is composed primarily of living cells—**sieve cells** in conifers, and **sieve tube elements** and **companion cells** in hardwoods. And unlike old, nonfunctioning xylem, which constitutes the wood of the tree and contributes to diameter increase, old phloem becomes crushed and may be reabsorbed into the tree or incorporated into the bark.

In addition to the **axial transport** system of the phloem and xylem, which transports materials longitudinally, trees also have transport cells arranged in **rays**. Rays are made up of parenchyma cells that grow radially in small layers that extend across the annual rings of phloem and xylem. Rays transport sugars and other compounds through the trunk, store starch, and assist in restricting decay in wood tissue.

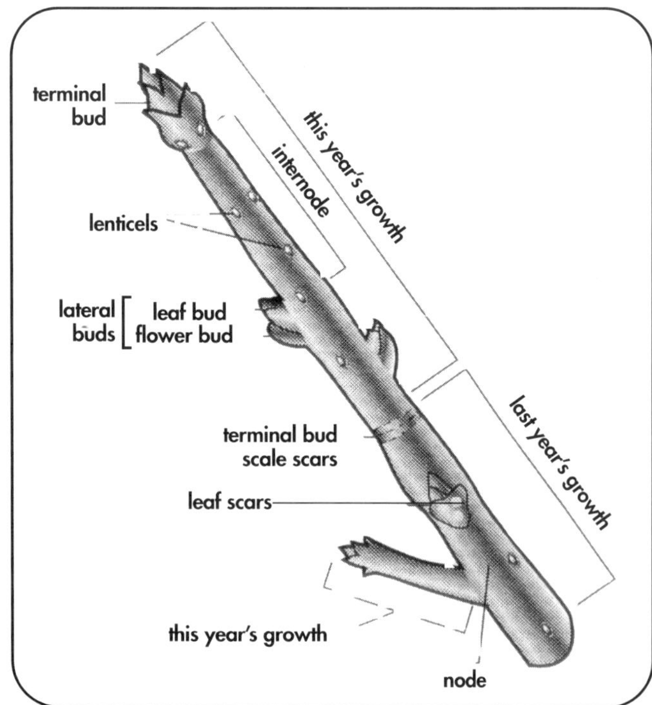

Figure 1.7 Twig anatomy showing twig extension growth.

Bark is the outer covering of a tree's branches and stems. It is a protective tissue that moderates the temperature inside the stem, offers defense against injury, and reduces water loss. Outer bark is composed of nonfunctional phloem and corky tissues. The cell walls are impregnated with wax and oil that minimize water loss. **Lenticels**, small openings in the bark, permit gas exchange. Many types of bark develop in trees. Beech (*Fagus)* trees have very smooth bark with little corky material, whereas cork oak (*Quercus suber*) produces thick layers of cork, which is made into stoppers for wine bottles.

Stems

Twigs are small stems that provide support structure for leaves, flowers, and fruit. Branches support twigs, and the trunk supports the entire crown. Buds can occur along the twig, at the base of each leaf, just under the bark, or at the tip of each twig.

Buds located at the end of a shoot are called the **terminal,** or **apical, buds**. Buds that occur along the stem are called lateral, or **axillary, buds** (Figure 1.7). Normally, the terminal bud is the most active on each branch or twig. Axillary, or lateral, buds are often **dormant**. Their growth may be inhibited by the **apical dominance** of the terminal bud, whereby the terminal bud inhibits the growth and development of laterals on the same shoot. When pruning removes terminal buds, dormant buds near the cut may be released, leading to new shoot development.

Adventitious buds are produced along stems or roots where primary meristems are not normally found. Their development may be stimulated by the loss of normal buds and the growth regulators they produce. Some tree species grow in groups in which individual trees have developed from adventitious buds on the roots, resulting in many trees having a common root system. Dormant buds are suppressed within growing tissues, originating with the first year's shoot and remaining suppressed beneath the bark until growth is triggered. When dormant buds elongate and produce shoots, these shoots are termed **epicormic**.

A **node** is a slightly enlarged portion of the twig where leaves and buds arise. The **internode** is the area between the nodes. Leaf scars and terminal bud scale scars are visible on new twigs and are useful in measuring annual twig elongation. On the outer surface of the twig, lenticels permit the exchange of gases.

Each branch of the tree is autonomous, producing and storing enough carbohydrates to sustain itself and exporting some to the trunk and roots. Branches

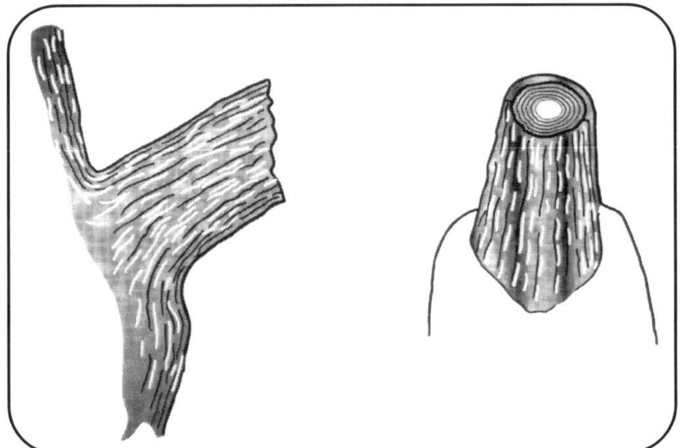

Figure 1.8 Branch collar. Trunk tissues grow around the branch.

generally do not import sugar from other parts of the tree. Each branch is similar in structure and function to the entire tree crown, yet branches are not simply outgrowths of the trunk. Instead, branches and trunks have a unique attachment form that is critical to the application of arboricultural practices such as pruning. Branches are strongly attached to the wood and bark beneath the branch but weakly attached to the wood and bark above the branch.

The annual production of layers of tissue at the junction of the branch to the stem is apparent, forming a shoulder or bulge around the branch base called the **branch collar** (Figure 1.8). In the crotch, the branch and trunk expand against each other. As a result, bark is pushed up to form the **branch bark ridge**. If bark in the crotch is surrounded by wood, it is called **included bark**. Included bark further weakens the crotch because the normal branch-to-trunk attachment is not formed.

Leaves

Leaves are the food producers of the tree. Leaves have cells with **chloroplasts** that contain a green pigment called **chlorophyll**. Chlorophyll is the primary leaf pigment that absorbs sunlight. The energy of the sunlight is trapped in the chloroplasts where it is converted to chemical energy in the form of sugar. This reaction is called **photosynthesis** (Figure 1.9).

A second role of leaves is using and controlling **transpiration**. Transpiration is the loss of water through the foliage in the form of water vapor, which helps cool the leaf and draw water up through the xylem. The structure of leaves is uniquely adapted to carry out the roles of photosynthesis and transpiration. Leaf blades provide a large surface area for the absorption of sunlight and carbon dioxide needed for photosynthesis. Because leaves are thin, no cells are far from the surface. This structure facilitates the exchange of gases and absorption of light.

The outer surface of a leaf is covered by a waxy layer called the **cuticle**. The cuticle functions to minimize desiccation (drying out) of the leaf. **Stomata**, small openings in the leaf surface, control the loss of water vapor and the exchange of gases. Carbon dioxide is absorbed into the leaf, while oxygen and water vapor are released. **Guard cells** regulate the opening and closing of the stomata in response to environmental stimuli such as light, temperature, and humidity (Figure 1.10).

Leaves have a network of conducting tissues comprising the veins, or vascular bundles. These veins are composed of both phloem and xylem tissue. They transport water and essential elements, and they carry food produced in the leaf cells to other parts of the tree.

Trees that shed their leaves every year are called **deciduous**. Trees that hold their leaves for more than one year are called **evergreen**. Deciduous trees generally lose their leaves in the fall as a result of cell changes and growth regulators that combine to form an **abscission zone** at the base of the leaf stalk or

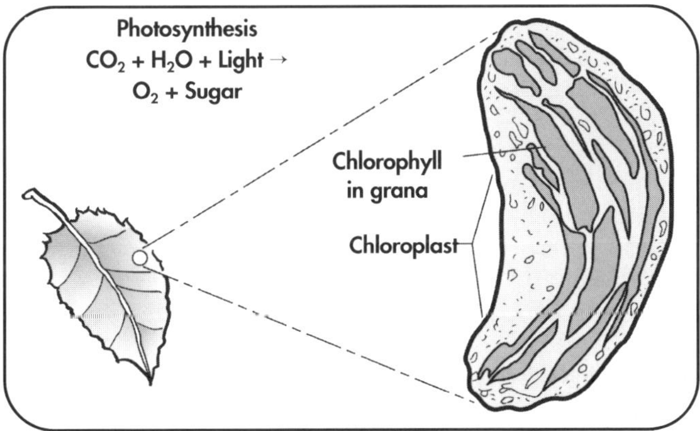

Figure 1.9 The light-absorbing, green pigment, chlorophyll, is located in chloroplasts within some leaf cells.

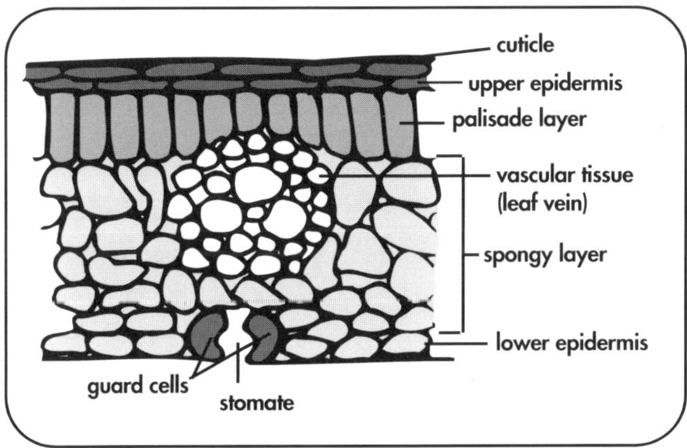

Figure 1.10 Cross section of a leaf blade.

petiole. The abscission zone has two functions: 1. to enable leaf drop in the fall, and 2. to protect the region of the stem from which the leaf has fallen against desiccation and pathogen entry.

Fall foliage color in deciduous trees results from the breakdown of chlorophyll and the expression of other pigments contained in the leaf. Short, sunny days combined with cold nights enhance the accumulation of sugars and trigger a decrease in chlorophyll production. This process allows other pigments, including **anthocyanins** (reds and purples) and **carotenoids** (yellows, oranges, and reds), to be unmasked.

Roots

The roots of trees serve four primary functions: anchorage, storage, absorption, and conduction. Larger roots are similar to the trunk and branches in structure. The main functions of large roots are anchorage, storage, and conduction. **Absorbing roots** are the small, fibrous, primary tissues that grow at the ends of the main, woody roots. The absorbing roots have epidermal cells that may be modified into root hairs, which aid in the uptake of water and minerals. As with shoot tips, root tips contain a meristematic zone where the cells divide and grow in length (Figure 1.11).

Roots grow where moisture and oxygen are available. Most absorbing roots are found in the upper 12 inches of soil. Horizontal, **lateral roots** are also usually near the soil surface. **Sinker roots** grow vertically downward off the lateral roots, providing anchorage and increasing the depth of soil exploited by the root system. These sinkers are usually found within a few feet of the trunk. The downward-growing **tap root** of young trees is usually choked out by expansion of roots around it or is diverted from its downward growth by unfavorable growing conditions. Roots may extend laterally for considerable distances, depending on the tree and soil conditions. Roots of trees grown in the open often extend two to three times the radius of the crown. The extent and direction of root growth is more a function of environment than genetics (Figure 1.12)

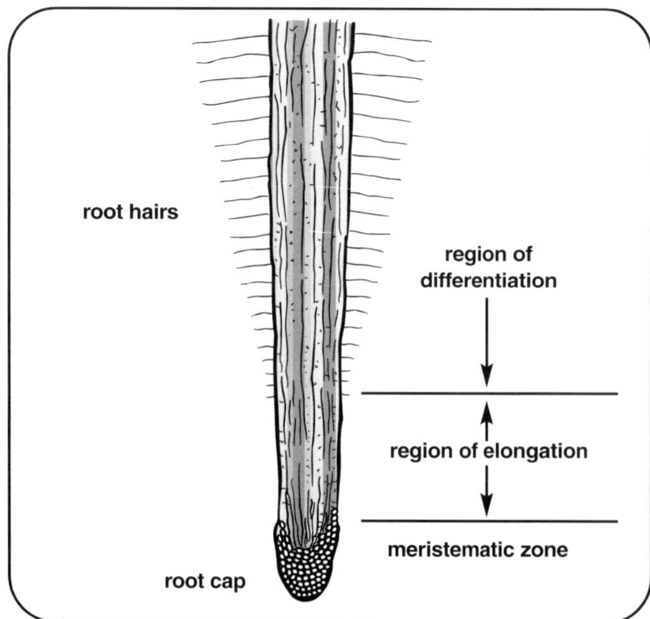

Figure 1.11 Root tip anatomy.

Many roots live in a symbiotic relationship with certain fungi. The result of the association is termed **mycorrhizae** (fungus roots) (Figure 1.13). In **symbiosis,** both organisms (the tree and the fungus in this case) benefit from the living arrangement. The fungi derive nourishment from the roots of the tree. In turn, the fungi aid the roots in the absorption of water and essential mineral elements.

TREE PHYSIOLOGY

Photosynthesis

Photosynthesis is the process by which green plants use light energy to build carbon molecules such as sugar. Literally, photosynthesis means "putting together with light." Photosynthesis takes place within cells that contain chloroplasts. Chloroplasts contain molecules of chlorophyll, the light-absorbing pigment that gives plants their green color. The raw materials necessary for photosynthesis are carbon dioxide and water. The tree absorbs carbon dioxide from the atmosphere through the stomata in the leaves. Light energy is absorbed and trapped in the chloroplasts. The light energy is converted to chemical energy and stored in the form of sugars and starches. Oxygen, the byproduct of photosynthesis, is released through the stomata.

The sugar products of photosynthesis are sometimes referred to as **photosynthate** or **carbohydrate**. Photosynthates are the building blocks for many other compounds required by the plant (Figure 1.14). Proteins, starch, fat, growth regulators, amino acids, and other important compounds are produced from photosynthate when combined with other essential elements such as nitrogen, potassium, sulfur, and iron. Much of the photosynthate is stored by the tree in the form of starch for later energy requirements.

Respiration

Respiration is the process by which the chemical energy generated by photosynthesis, and stored as starch or sugar, is used by the tree. Sugars and starch

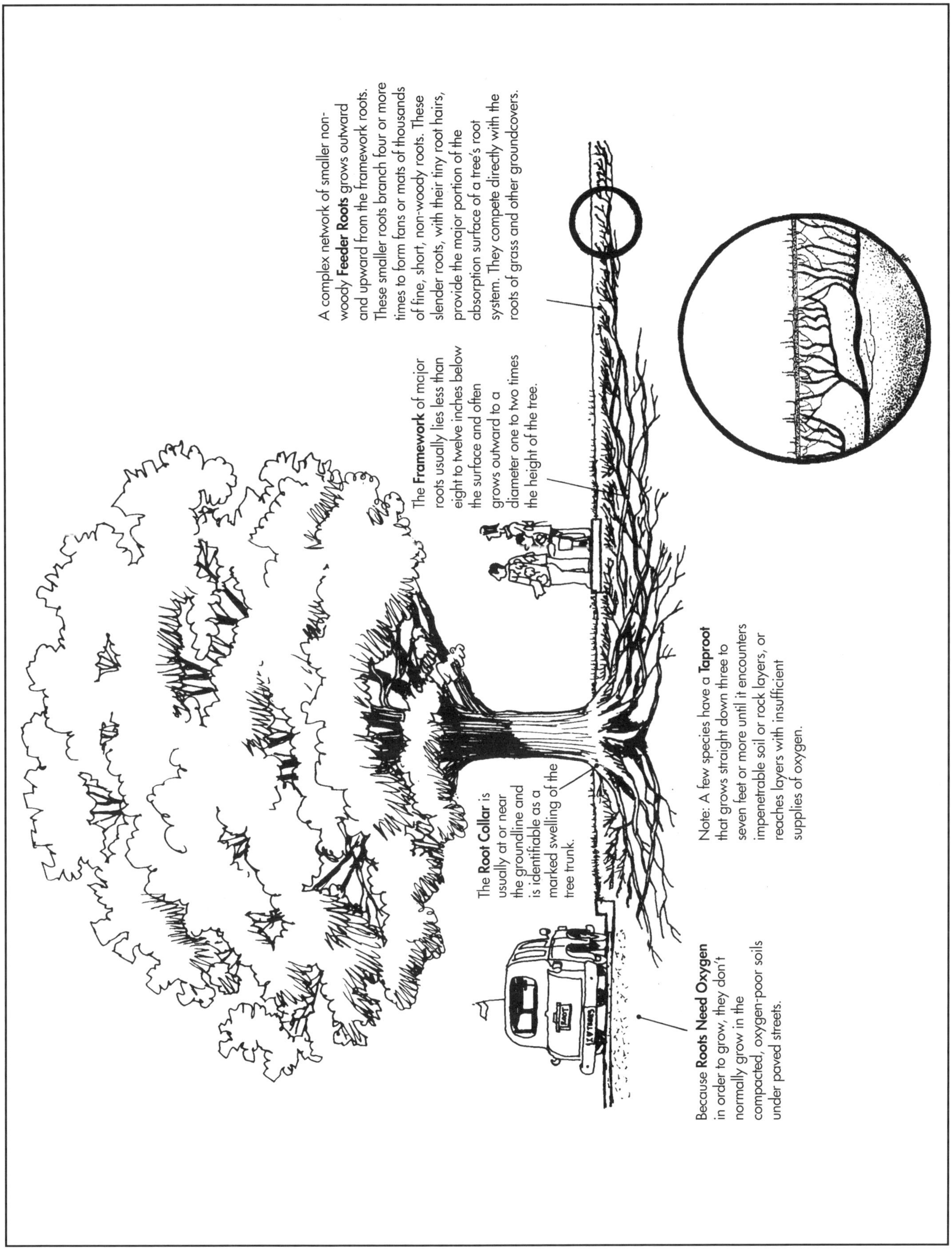

A complex network of smaller non-woody **Feeder Roots** grows outward and upward from the framework roots. These smaller roots branch four or more times to form fans or mats of thousands of fine, short, non-woody roots. These slender roots, with their tiny root hairs, provide the major portion of the absorption surface of a tree's root system. They compete directly with the roots of grass and other groundcovers.

The **Framework** of major roots usually lies less than eight to twelve inches below the surface and often grows outward to a diameter one to two times the height of the tree.

The **Root Collar** is usually at or near the groundline and is identifiable as a marked swelling of the tree trunk.

Because **Roots Need Oxygen** in order to grow, they don't normally grow in the compacted, oxygen-poor soils under paved streets.

Note: A few species have a **Taproot** that grows straight down three to seven feet or more until it encounters impenetrable soil or rock layers, or reaches layers with insufficient supplies of oxygen.

Figure 1.12 Roots grow where water, oxygen, and space are available.

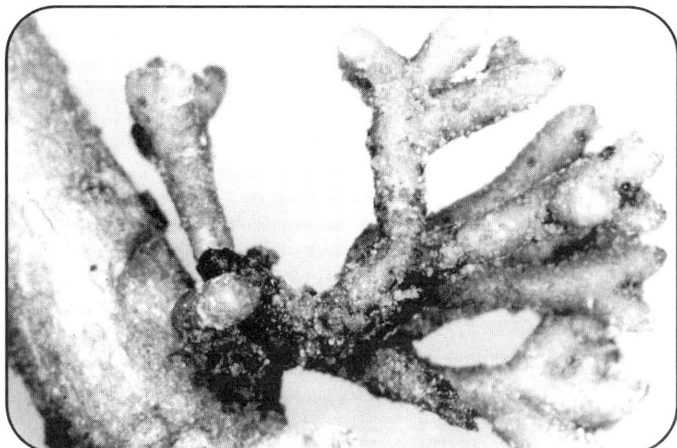

Figure 1.13 One of many types of mycorrhizae found on or in roots. Mycorrhizae facilitate absorption of elements.

are actually chains of carbon, hydrogen, and oxygen molecules chemically bonded together. When the bonds are broken, energy is released, and carbon dioxide and water are given off. The tree uses the energy released for all biological functions.

Both plants and animals respire. Respiration is a constant process of converting food into energy. Plants produce their own food, however. Thus, it is important that overall photosynthesis (food making) exceeds respiration (food using). When respiration takes place in the absence of photosynthesis, the tree must rely on stored energy reserves. If this occurs over a long period of time, the tree will eventually run out of energy and die. A practical example is a tree that is repeatedly defoliated. Without foliage, photosynthesis stops, and the tree cannot produce and store starches. Yet as long as the tree is living, respiration takes place, and energy is consumed.

Oxygen is required for normal, aerobic respiration. Trees are not capable of respiration in the absence of free oxygen (anaerobic) for long. Under flooded conditions or with severe soil compaction, oxygen is scarce and respiration uses food at a much higher than normal rate. If the conditions persist, the roots will die.

Transpiration

Transpiration is the loss of water from leaf surfaces in the form of water vapor. The evaporation of water cools the leaves and creates a "transpirational pull" that moves water up through the xylem. The cuticle helps prevent uncontrolled water loss from epidermal cells on the leaf surface. Gas exchange, where water vapor and oxygen are released and carbon dioxide is absorbed, takes place through stomata, which are located primarily on the underside of the leaves. Each stomatal pore is lined by two guard cells that

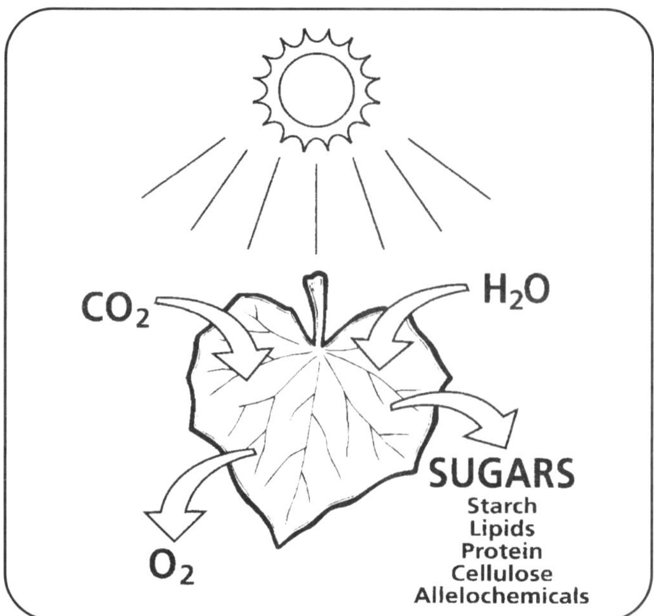

Figure 1.14 Biochemical building blocks from which plants are constructed are manufactured from the simple sugars produced during photosynthesis.

regulate the opening and closing of the stomate. Stomatal opening is influenced by environmental conditions such as light, temperature, and humidity. Stomates usually open in the light and close in the dark.

Temperature, humidity, and available water affect the rate of transpiration. Transpirational water loss is also affected by anatomical features such as cuticle thickness, the presence of hairs on the leaf surface, and the number and location of stomata. Some plants with a thick cuticle, small leaves, and sunken stomata are adapted to hot and dry conditions.

Arborists sometimes use **antitranspirant** sprays to artificially close stomatal pores or to add an impervious coating on the leaf surface, thus reducing water loss during transplanting or under drought conditions. Use of these materials, however, reduces evaporative cooling of leaves, reduces uptake of carbon dioxide, and can reduce the rate of photosynthesis.

Absorption, Translocation, and the Vascular System

Water is essential for all living cells. Most biological reactions, including photosynthesis, require water. Water maintains cell turgidity and is necessary to transport essential elements within the xylem. Water and mineral elements are absorbed from the soil by the roots. The tree uses some of this water for growth and metabolism, but most is lost through transpiration. This water loss creates the "transpirational pull" that helps move water through the xylem. The xylem can be thought of as a continuous column of water, where the evaporation of water molecules from the leaves pulls water up through the tree.

Water enters young roots or mycorrhizal roots by a process called **osmosis**. Osmosis is the movement of water through a membrane from a region of high water potential (concentration) to a region of low water potential. Pure water has the highest potential; adding anything such as minerals or sugar lowers the potential. Water normally moves into roots where the water potential is lower than in the surrounding soil. If the water potential is lower in the soil than the root cells, water will actually move out of the roots into the soil. An example is when salt concentrations are high in the soil, such as from deicing or excessive fertilizer application.

Phloem transport is the movement of photosynthate and other compounds in the phloem. The phloem is composed of sieve cells in gymnosperms and sieve tube elements in angiosperms. The carbohydrate products of photosynthesis are actively pumped through the phloem, a process that requires energy.

In discussing phloem transport, the terms **source** and **sink** are often used. Leaves are the source of photosynthate. The photosynthate moves through the phloem in a direction from source to sink, from areas high in sugar concentration to areas where more is required. Sinks are plant parts that use more energy than they produce. Almost all plant parts, including young leaves, are sinks at some time. It is sometimes thought that sugars are produced in the leaves and transported, through the phloem, exclusively to the roots for storage. Most photosynthate is either utilized or stored in proximity to where it is manufactured, although it actually can move in either direction in the phloem.

The movement of water in the xylem and photosynthate in the phloem are examples of longitudinal, or axial, transport. **Radial transport** is the horizontal movement of water or nutrients within the tree between cells of different ages, primarily through ray cells. Rays are living channels of cells through which water, elements, and carbohydrates move laterally.

Control of Growth and Development

The growth and development of a tree are the results of the interaction between its genetic potential and the surrounding environmental conditions. There are many examples of this interaction. In nature, sweetgum *(Liquidambar styraciflua)* may attain a height of 150 to175 feet. Yet rarely does the species grow that large in urban areas. While sweetgum may have the genetic potential to grow 150 feet tall, the urban environment limits the expression of that potential. Other genetic components work similarly. The range of size, fall color, and form seen in many cultivars of red maple *(Acer rubrum)* represent the breadth of genetic potential within the species. Yet in red maple, 'Red Sunset' will never look like 'Armstrong' even when grown under the same environmental conditions because the genetic makeup of these two cultivars is different.

Plant systems, like all living organisms, respond to environmental stimuli (Figure 1.15). Developmental responses to light, gravity, and temperature can be essential to the survival of a tree. For example, a long period of exposure to cold may be necessary to induce budbreak, flowering, or seed germination. The coordination of processes in trees is controlled in part by plant growth substances. Plant growth regulators are naturally occurring compounds that act in small quantities to regulate plant growth and development. The major plant hormone groups include auxins, gibberellins, cytokinins, ethylene, and abscisic acid. These plant growth substances work in concert to control such functions as cell division, cell

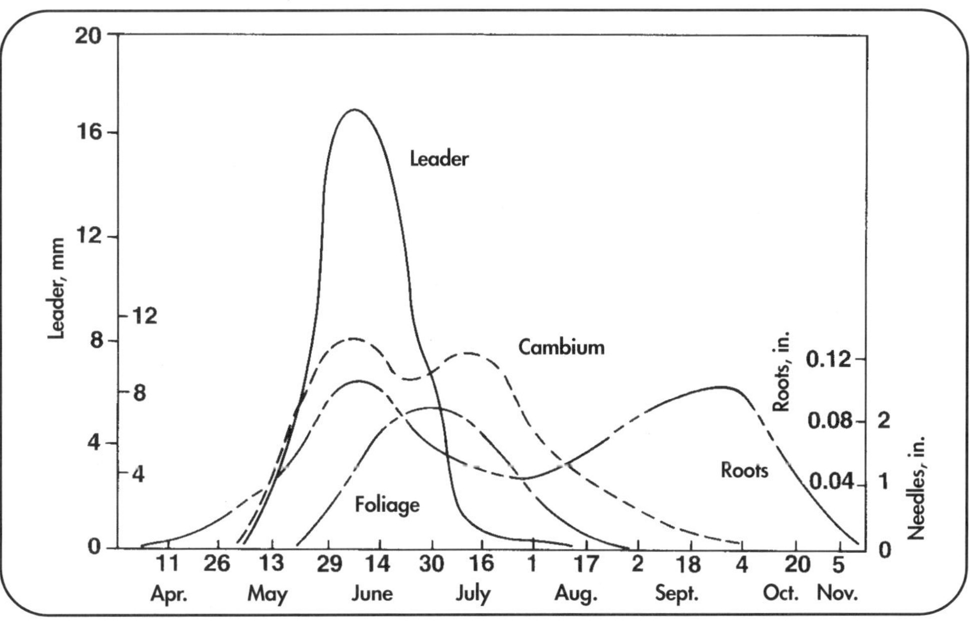

Figure 1.15 Typical seasonal growth of a pine in a temperate climate. Note the late-season growth activity.

elongation, flowering, fruit ripening, leaf drop, dormancy, and root development.

Auxin is a plant growth substance that has been linked to several developmental phenomena. Although auxin is produced primarily in shoot tips, it is known to be important in root development. Synthetic forms of auxin are sold commercially to enhance the rooting of cuttings. The fact that auxin is produced in the shoot tips may partially explain why heavy crown pruning to compensate for root loss during transplant has not proved to be effective. Conversely, **cytokinin**, produced in the roots, is instrumental in shoot initiation and growth. Auxin and cytokinin exist in a delicate balance in plants, regulating shoot and root growth.

Auxin has also been found to be involved in **tropisms**. A tropism is the orientation of the direction of growth in response to an external stimulus. An example of a tropism is the upward orientation of stem growth or the downward direction of root growth. These phenomena are responses to gravity and are examples of **geotropism**. Light also has a strong influence on the direction of plant growth. A tree that grows at an angle toward the sunlight is exhibiting **phototropism** (Figure 1.16).

Apical dominance is also a result of plant growth regulators. Growth regulators present in terminal buds inhibit the growth and development of lateral buds on the same shoot. Strong dominance is confined primarily to the current season's shoot growth. During the following season, lateral buds start growing. If the new lateral shoots outgrow the original terminal shoot year after year, a round-headed, or **decurrent**, tree will result. **Excurrent** trees tend to have strong apical control, resulting in upright trees with strong central leaders (Figure 1.17). Although some species tend toward one of these forms throughout their lives, all trees start out with excurrent traits as juveniles and most become more decurrent as they mature. Sweetgum *(Liquidambar styraciflua)*, tuliptree *(Liriodendron tulipifera)*, and most conifers are excurrent forms. In decurrent trees, the ability to maintain the strong central leader is lost, and the rounded form develops.

Figure 1.16 Phototropism. Plants grow toward light due to differential distribution of auxin.

Maturity and environmental conditions can also influence the growth form of trees. Many pines maintain a strong excurrent form throughout most of their lives but develop a rounded top as they age. Limited access to sunlight, such as in forest understory conditions, can cause a normally rounded tree to grow upright.

It is important to take a holistic view of tree growth and development. A delicate and changing balance of chemical signals controls the metabolic processes of photosynthesis, respiration, and all biological functions within a tree. Tree growth regulators remain in a dynamic equilibrium. Environmental stimuli, including those caused by humans, trigger changes in concentrations that, in turn, stimulate changes in resource allocation, which results in a response in the tree's growth or development.

A System of Defense

Trees cannot actively fight or move away from harm, but that does not leave them defenseless. Trees have a number of features that serve as protection: thick bark, thorns, leaf hairs, thick cuticles, and many others. In addition, certain cellular materials may resist decay or may be indigestible by insects. Another defense mechanism is the production of chemicals that resist insect feeding, pathogen infection, or decay.

A developmental process unique to trees is the ability to compartmentalize or "wall off" decay.

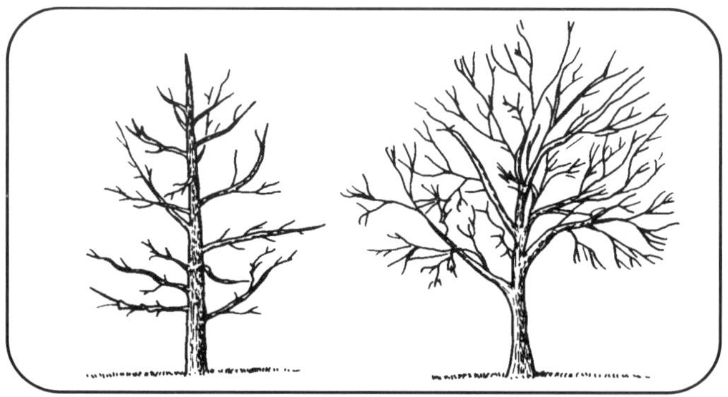

Figure 1.17 Excurrent tree. Decurrent tree.

Compartmentalization is the process by which trees limit the spread of discoloration and decay. After a tree has been wounded, reactions are triggered that cause the tree to form boundaries around the wounded area.

Alex Shigo has proposed a model of this compartmentalization process called **CODIT** (Compartmentalization Of Decay In Trees) (Figure 1.18). In Shigo's model, the tree forms four barrier "walls." Wall 1 resists vertical spread by plugging xylem vessels. Wall 2 resists inward spread by the more compact latewood cells and by depositing chemicals in these cells. Wall 3 inhibits lateral spread by activating ray cells to resist decay. These three walls form the **reaction zone**. Wall 4 is the next layer of wood to form after injury, and it protects against the outward spread of decay. This is the barrier zone. Wall 1 is the weakest, and Wall 4 is the strongest barrier. At times, the tree cannot resist the spread of aggressive pathogens. It is fairly common for walls 1, 2, and 3 to fail, allowing decay to spread inside the tree, forming a hollow cavity. Wall 4 rarely fails, except where canker-causing fungi restrict its development or kill the cambium (Figure 1.19). The barrier zone is strong chemically but weak structurally. The process that resists the spread of disease can also lead to **shakes** and cracks.

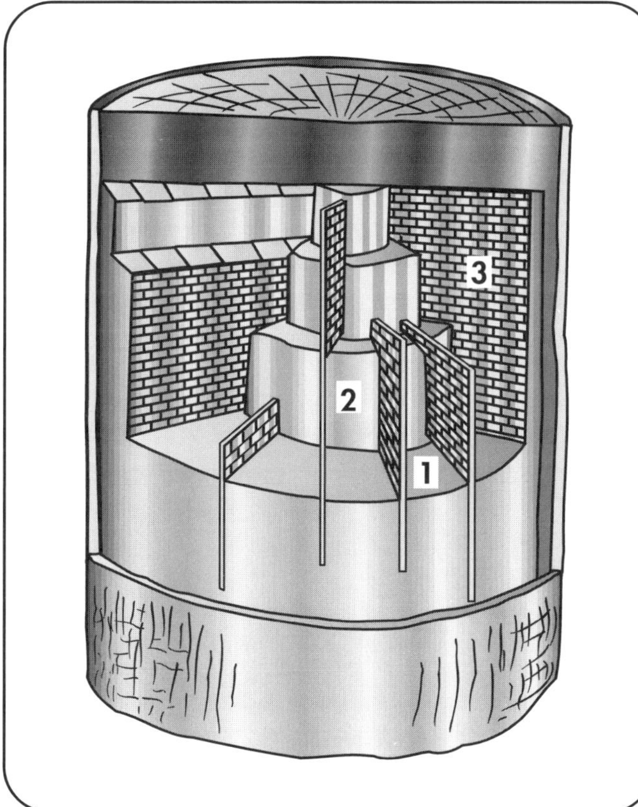

Figure 1.18 CODIT. Wall 1 is formed when the tree responds to wounding by "plugging" the upper and lower vascular elements to limit vertical spread of decay. Wall 2 is formed by the last cells of the growth ring, limiting inward spread. Wall 3 is composed of ray cells that compartmentalize decay by limiting lateral spread. Wall 4 (not shown), the strongest wall, is the new growth ring that forms after injury.

PALMS

Palms are monocots and have more in common biologically with grasses than with hardwood trees. Palms do not have a cambium layer or growth rings of xylem. Instead, they have vascular bundles of phloem and xylem interspersed within the stem. The stem develops all the vascular bundles it requires for life behind the primary bud as it grows. The stem is also capable of storing starch in parenchyma cells. Photosynthesis takes place in the fronds, which dominate the palm crown. The reproductive structures (flowers and fruit) are also in the crown.

The root system of palms is also very different from that of hardwood trees. The roots lack secondary growth (cambium meristem) and, other than the primary seedling root, they are adventitious—developing from stem internodes. There are many species of palms, and they vary greatly in structure and form. This text provides merely a basic contrast in the biology of palm compared to dicotyledenous trees.

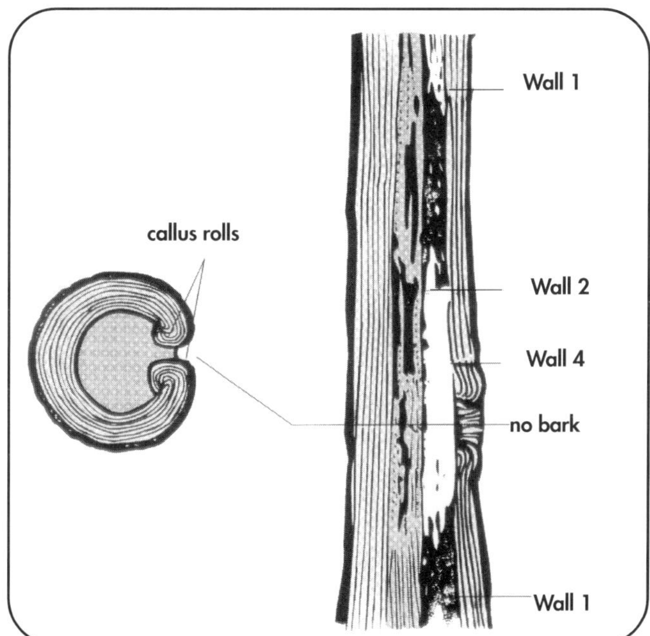

Figure 1.19 Compartmentalization of decay. Wall 4 prevents decay from entering new wood. Wall 3 (not shown) and Wall 2 have failed to prevent the decay from spreading laterally and internally.

Chapter 1 Workbook

1. Sites of rapid cell division in the shoot tips, root tips, and cambium are called ________________.
2. Buds are covered by ______________ that leave permanent scars on the twig when the buds break dormancy. These scars can be useful in tree identification.
3. The tendency for terminal buds to inhibit the growth of lateral buds is called __________ ________________.
4. The "food factories" of trees are the _____________.
5. The process of ______________________ combines carbon dioxide and water in a reaction driven by light to produce sugars. _______________ is also a product of this reaction.
6. The green color of leaves is created by the presence of __________________, which is necessary for photosynthesis to take place.
7. ____________________ is the loss of water vapor from the leaves.
8. The opening and closing of ______________ allows for gas exchange, and transpiration is controlled by the _________ ________.
9. Water and dissolved essential minerals are transported within the tree in the _____________. The ______________ conducts carbohydrates.
10. The _________________ is a layer of meristematic cells located between the phloem and the xylem.
11. The _____________ __________ is formed when trunk tissue grows around branch tissues. As the branch and trunk tissues expand against each other in the crotch, the ___________ ______ ___________ is formed.
12. __________ protects the branches and trunk of a tree from mechanical injury and desiccation.
13. Name four functions of the root system.

    a.

    b.

    c.

    d.
14. Water enters young roots or mycorrhizal roots by a process called _________________.
15. The orientation of growth in response to an external stimulus is called ________________. Two examples are _______________ and _____________________.
16. CODIT stands for

    C__________________

    O____

    D_________

    I____

    T_________
17. Trees with upright growth and a strong, central leader are said to exhibit _______________ growth. More rounded trees, which are often broader than they are tall, have _______________ growth habits.

18. Roots and fungi called ________________ exist in a symbiotic relationship, which aids in the uptake of water and minerals.
19. The process by which chemical energy, stored as sugar and starch, is released is called ________________.
20. Trees that lose their leaves in the fall are called ________________. Trees that maintain their leaves for more than one year are called ________________.
21. Label the following diagrams:

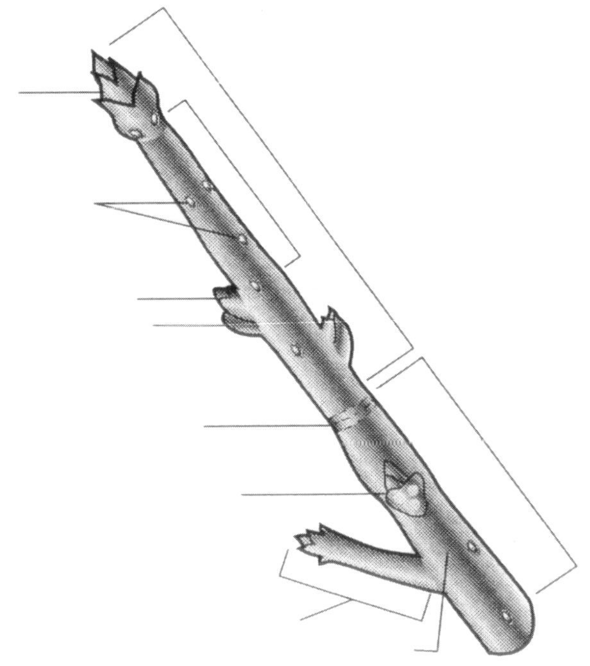

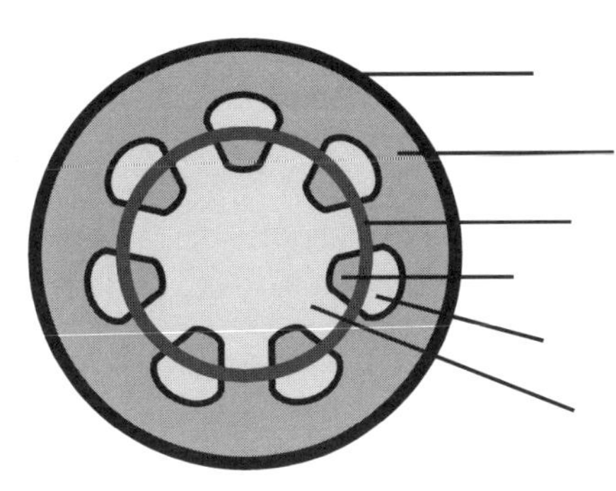

MATCHING

___ auxin	A. uses more energy than it produces
___ chlorophyll	B. mostly located in the upper 12 inches of soil
___ cuticle	C. "stalk" of a leaf
___ petiole	D. cells that cross the phloem and xylem for radial transport
___ internode	E. waxy covering of a leaf
___ lenticel	F. small openings in stems for gas exchange
___ ray	G. plant hormone
___ absorbing roots	H. between the nodes of a twig
___ source	I. mature, green leaves—sugar producers
___ sink	J. green pigment

CHALLENGE QUESTIONS

1. Describe how growth rings are formed. Are they truly annual? What information can be obtained by examining growth rings? How might the rings be useful in diagnosis?

2. What is the relationship between photosynthesis and respiration? What is net photosynthesis? What might be the effect on a tree if respiration exceeds photosynthesis for an extended period of time? How could this occur?

3. Explain the process of compartmentalization of decay in trees. What are the four walls, and how do they function? How can trees be hollow and still remain vigorous?

4. Does a branch higher in a tree receive direct nourishment from lower branches? Discuss the vascular structure of trees and how it relates to branch attachment.

5. *Armillaria mellea* has been described as an "opportunistic" fungus that functions both as a mycorrhiza and a disease. How might this be explained?

SAMPLE TEST QUESTIONS

1. When cutting through a tree with a chain saw or drilling into a tree, you would pass through (in order)
   a. bark, cambium, phloem, xylem
   b. bark, phloem, cambium, xylem
   c. bark, cambium, xylem, phloem
   d. bark, xylem, phloem, cambium
2. If the terminal bud is removed in pruning
   a. growth may be stimulated in lateral buds
   b. flowering is stimulated to enhance fruit production
   c. the branch will die back
   d. all of the above
3. The growth rings of a tree
   a. are visible because of the rapid growth rate of earlywood relative to latewood

b. can be counted to approximate a tree's age
c. can give information about growing conditions in previous years
d. all of the above

4. Which layer of cells is responsible for outward growth and increased girth of a tree?
a. cambium
b. pith
c. epidermis
d. cortex

5. Mycorrhizae are
a. collar-rot fungi
b. elongated underground stems producing sucker sprouts
c. a symbiotic relationship between fungi and roots
d. cells in which photosynthesis takes place

Other Sources of Information

(See pages v–vi for complete bibliographic information.)

Harris et al., 1999. *Arboriculture: Integrated Management of Landscape Trees, Shrubs, and Vines.*
Kramer and Kozlowski, 1979. *Physiology of Woody Plants.*
Shigo, 1986. *A New Tree Biology.*

CHAPTER 2
TREE IDENTIFICATION

CHAPTER 1 TREE BIOLOGY

CHAPTER 2 TREE IDENTIFICATION

CHAPTER 3 TREE/SOIL RELATIONS

CHAPTER 4 WATER MANAGEMENT

CHAPTER 5 TREE NUTRITION AND FERTILIZATION

CHAPTER 6 TREE SELECTION

CHAPTER 7 INSTALLATION AND ESTABLISHMENT

CHAPTER 8 PRUNING

CHAPTER 9 TREE SUPPORT AND PROTECTION SYSTEMS

CHAPTER 10 DIAGNOSIS AND PLANT DISORDERS

CHAPTER 11 PLANT HEALTH CARE

CHAPTER 12 TREE ASSESSMENT AND RISK MANAGEMENT

CHAPTER 13 TREES AND CONSTRUCTION

CHAPTER 14 SAFETY

CHAPTER 15 CLIMBING AND WORKING IN TREES

CHAPTER 2 TREE IDENTIFICATION

Objectives

1. Understand how all plants are classified and how scientific names are based on the classification system.
2. Explain what a scientific name is, why scientific names are used, and how they are written.
3. Explain how plant characteristics such as growth habit, texture, and color can be used in tree identification.
4. Describe how leaf arrangement is used to help identify trees.
5. Become familiar with the various leaf shapes and types of leaf margins, bases, and apices.
6. Learn to use bud and twig characteristics to identify trees without leaves.

Key Terms

alternate	**family**	**leaf apex**	**nomenclature**	**serrate**
angiosperm	**foliage**	**leaf base**	**opposite**	**species**
class	**genus**	**leaf margin**	**order**	**specific epithet**
conifer	**gymnosperm**	**lobe**	**palmate**	**taxonomy**
cultivar	**identification key**	**monocotyledon**	**phylum**	**variety**
dicotyledons	**kingdom**	**morphology**	**pinnate**	**whorled**
division				

INTRODUCTION

Identification of tree species is usually the first step before prescribing tree care. Arborists should be able to identify a tree before attempting an accurate diagnosis or treatment recommendation. Plant species identification is a requirement in order to apply pesticides legally. Accurate identification requires a combination of knowledge and experience. Once identification skills have been learned, proficiency will come with practice and repeated exposure to woody landscape plants at different times during the year.

PLANT CLASSIFICATION

Plant classification, or **taxonomy**, is based on biological characteristics. The highest classification level is the **kingdom**, and, not too surprisingly, trees are in the plant kingdom. The second classification level is the **division** or **phylum**. This level further separates vascular plants (plants with xylem and phloem) from plants lacking vascular tissue. Vascular plants are subdivided into those with seeds covered by an ovary, **angiosperms**, and the **gymnosperms**, with "naked" seeds.

Angiosperms are flowering plants, which include most deciduous trees and broad-leaved evergreens. Conifers, or cone-bearing plants, are gymnosperms. Angiosperms are divided into two **classes**, the **dicotyledons** (dicots, two seed leaves) and the **monocotyledons** (monocots, one seed leaf). The dicots include most broadleaf tree species. Monocots include grasses, lilies, orchids, and palms. The vascular tissues of monocots are in bundles, scattered throughout the stem. Because these bundles do not increase in girth, palm stems have little ability to increase in diameter.

Classes of plants are separated into **orders** and then **families**. Plants in the same family have common characteristics, most notably their types of flowers and fruits. Honeylocust *(Gleditsia)* and redbud *(Cercis)* are both in the Fabaceae family (legumes). Their flowers are morphologically similar, and both bear their seeds in pods. Plants that are very closely related may be classified in the same **genus**. Because they are closely related, they show similar characteristics, particularly in their reproductive structures. For example, all oaks are in the genus *Quercus*. The **species** is the level that identifies the particular plant. The species name consists of the genus name and the **specific epithet** (Table 2.1).

Table 2.1. Classification of *Acer saccharum*, sugar maple.

Kingdom	Plantae
Phylum (Division)	Magnoliophyta (angiosperms)
Class	Magnoliopsida (dicotyledons)
Order	Sapindales
Family	Aceraceae
Genus	*Acer*
Specific epithet	*saccharum*

A general understanding of plant classification can help an arborist learn to identify plants. Plants that are closely related have similar characteristics. This knowledge can be helpful in diagnosis because trees in the same family are often susceptible to the same diseases and insect pests.

PLANT NOMENCLATURE

Plant **nomenclature** is the naming of plants. Arborists are often familiar with common names of trees because they have learned to identify them through years of field experience. It is important to realize, however, that the exclusive use of common names can lead to confusion and misunderstanding. One tree may have several common names. *Carpinus caroliniana* is known as American hornbeam, blue beech, ironwood, and musclewood. Two different tree species may have the same common name. For instance, *Magnolia* x *soulangiana, Spathodea campanulata,* and *Liriodendron tulipifera* are all called tuliptree in different areas of the world. Common names can even be misleading. For example, Douglas-fir is not a fir, baldcypress is not a cypress, and mountainash is not a species of ash.

Each plant has a unique scientific name that is the same throughout the world. Scientific names of plants are based on a species classification system, and each scientific name has at least two parts. The first part of a scientific name is the genus, which is written with a capitalized first letter. Plants in the same genus are closely related and show similar characteristics, particularly in their flowers and fruit. The second part identifies the specific epithet and is not capitalized. The genus and specific epithet together comprise the species. Some species are further divided into **varieties** or **cultivars**. A variety is a subdivision of a species that has a difference and breeds true to that difference. Variety names are not capitalized. Cultivars are cultivated varieties, and the names are written with single quotes with the first letter capitalized. Common names should not be capitalized unless they include a proper name such as "European." In writing, scientific names are either underlined or italicized. Examples of properly written scientific names are

Eucalyptus citriodora—lemon-scented gum

Gleditsia triacanthos inermis—
thornless honeylocust

Prunus cerasifera 'Atropurpurea'—
purpleleaf plum

PRINCIPLES OF IDENTIFICATION

Woody plant identification is based on **morphology**, which is the size, shape, and appearance of plant parts. A fundamental knowledge of woody plant anatomy is therefore essential. Although arborists usually concentrate on the leaves and overall form when identifying trees and shrubs, classification is often based more on the reproductive structures: the flowers and fruit. A good arborist learns to identify trees using many characteristics, including trunk form and texture, twigs, leaves, flowers, and fruit. This will enable identification in any season. In temperate regions, an arborist who learns tree identification by leaves alone will be able to identify deciduous trees half of the year only. Figures 2.1 through 2.4 show characteristics of leaves and their arrangement on a stem that can be used to help identify trees.

Many trees can be identified at a distance based upon their form and branching characteristics (growth habit). The American elm *(Ulmus americana)*, with its vase-shaped growth habit and overarching limbs, is unmistakable. Sugar maples *(Acer saccharum)* in the autumn can be picked out of a forest from miles away because of their brilliant fall colors. Live oak *(Quercus virginiana)* is commonly identified by its spreading habit and Spanish moss. In other instances, however, a group of arborists may find themselves

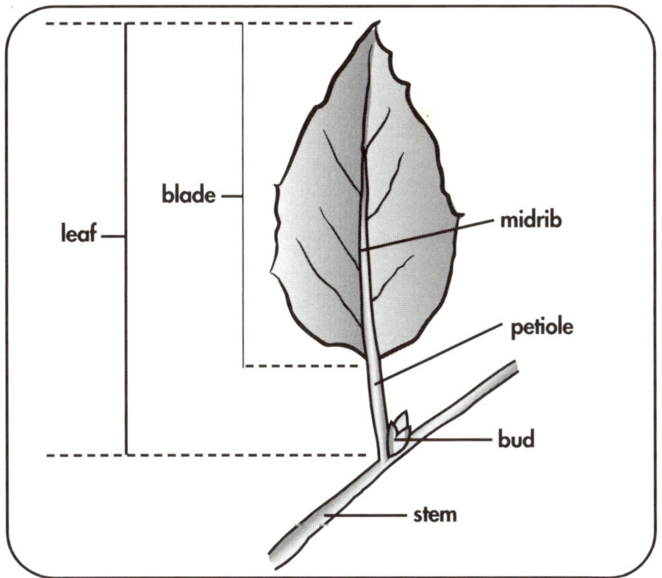

Figure 2.1 Anatomy of a simple leaf.

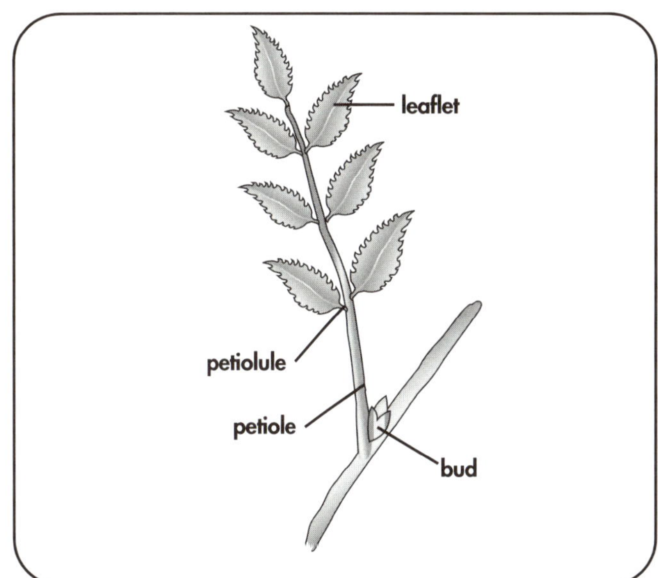

Figure 2.2 Compound leaf.

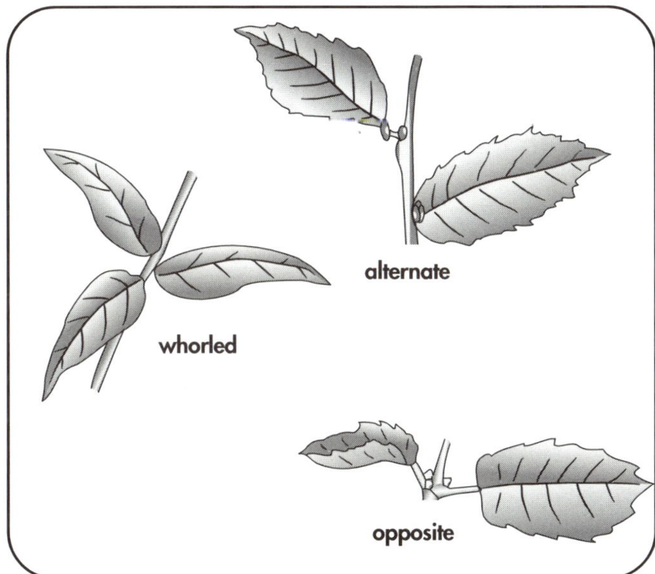

Figure 2.3 Leaf arrangements on stems.

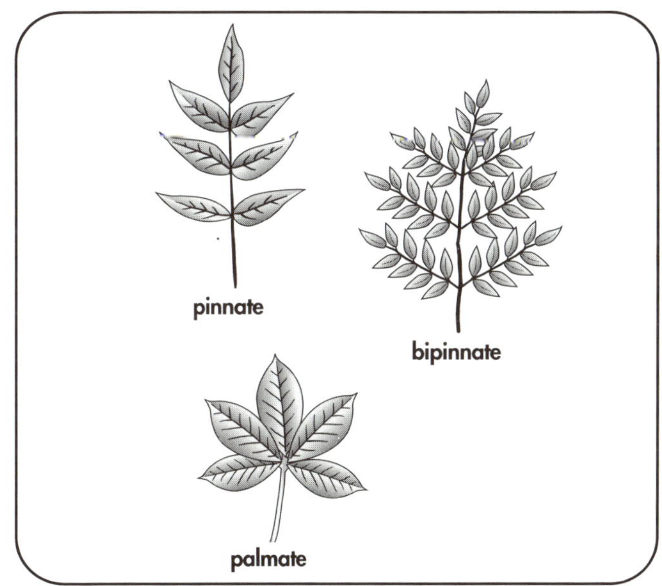

Figure 2.4 Arrangement of leaflets on compound leaves.

gathered around a plant, closely examining the details of a twig in order to identify it.

It is important to use all available information in tree identification. Identifying some tree species requires recognizing relatively minor leaf, bud, or twig characteristics. For example, the type of **leaf margin**, shape of the **leaf base**, presence of hairs on the upper or lower leaf surfaces, or the color of young twigs may be required to distinguish between two closely related species (Figures 2.5 through 2.7). For winter identification in temperate regions, arborists must be familiar with such characteristics as bark, branching habit, twigs, buds, and pith to identify certain trees (Figures 2.8 through 2.11). Sometimes tree identification is based upon more than visual senses. Smell, and even taste, may be useful to determine unique characteristics of twigs, leaves, flowers, or fruit. Arborists sometimes use little tricks to help them identify certain species. One such example helps narrow down the genus based on whether a tree has **opposite** or **alternate** leaf arrangement. In temperate zones, most of the trees with opposite leaf arrangement fall into four genera represented by "MAD Buck," for **m**aple, **a**sh, **d**ogwood, and **buck**eye.

There are also a few simple ways to distinguish between the major groups of **conifers,** or cone-bearing trees. Pines *(Pinus)* have needles usually borne in clusters of two, three, or five. Counting the

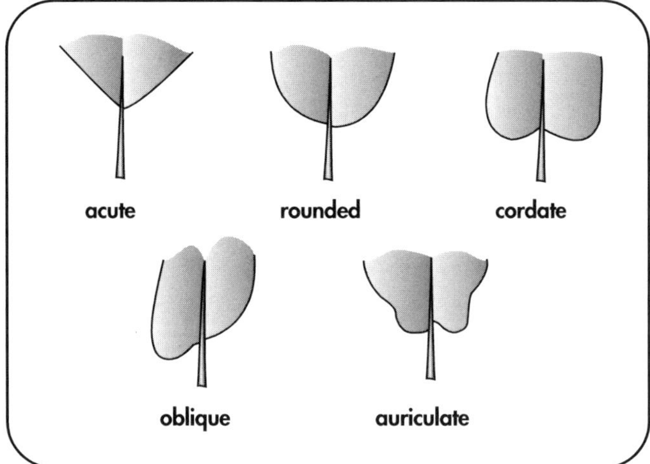

Figure 2.5 Leaf bases.

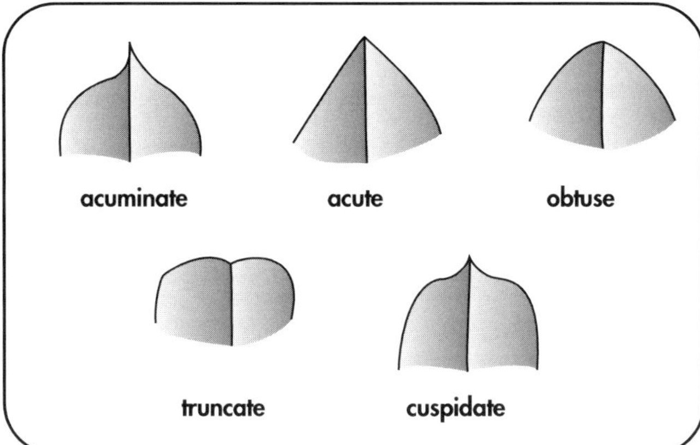

Figure 2.6 Leaf apices.

needles can help identify the species. Spruces (*Picea*) and firs (*Abies*) produce their needles singly. The needles of firs detach from the stem, leaving a circular "pad," whereas spruce needles leave a tiny stalk. Other conifers may have awl-like or scale-like **foliage** (Figure 2.12).

TERMS USED IN IDENTIFICATION

Many tree and shrub reference books contain **identification keys**. A key is a step-by-step method for unlocking the identity of a plant. Identification keys use terminology that describes the shape, texture, and arrangement of the leaves; buds; twig shape and texture; and the type of flowers and fruits. The more commonly used woody plant identification terms are illustrated or explained in this chapter.

An identification key may be used to systematically determine the identity of a plant. Most keys consist of a flow chart of "yes" or "no" questions. The user narrows down the possibilities by determining whether the leaf (bud) arrangement is **opposite** or **alternate**, whether the plant has simple or compound

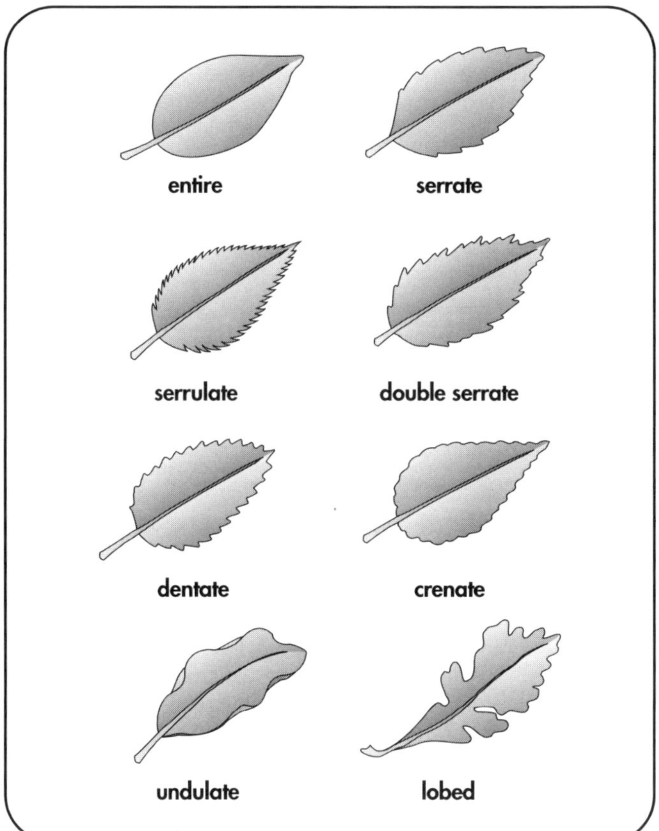

Figure 2.7 Leaf margins.

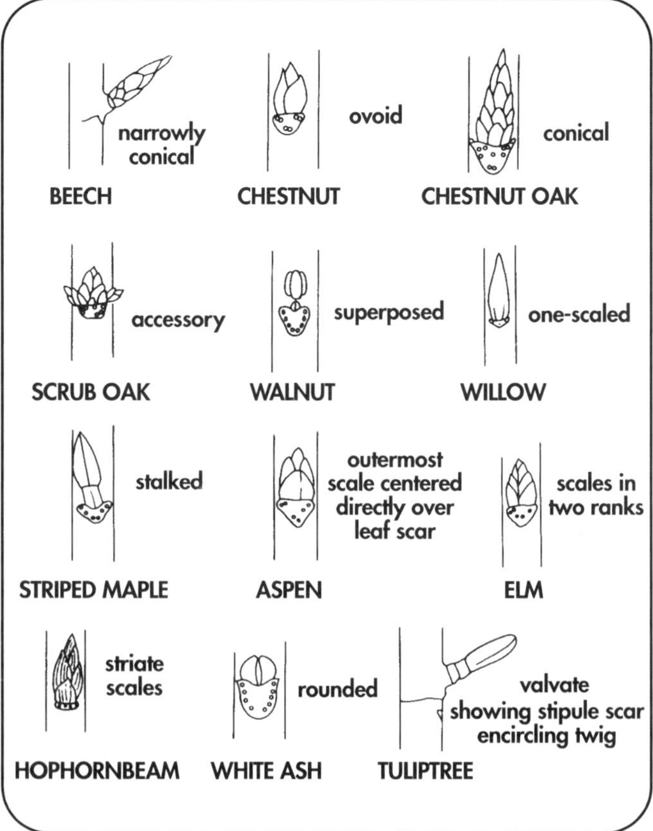

Figure 2.8 Various types of buds found on trees.

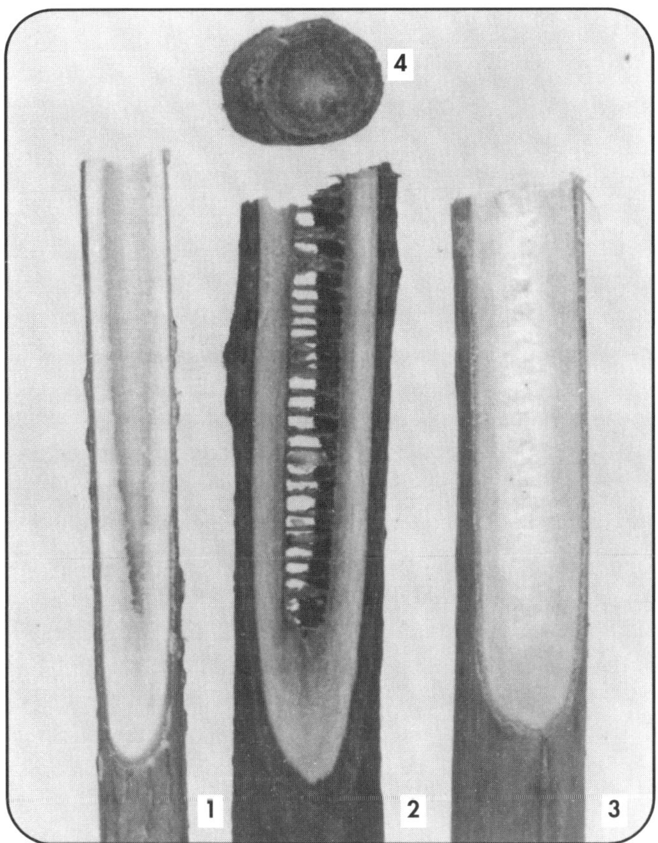

Figure 2.9 Pith characteristics: 1. weeping forsythia (hollow); 2. walnut (chambered); 3. winter honeysuckle; 4. star-shaped pith of oak.

Figure 2.10 Exfoliating bark of paperbark maple (*Acer griseum*).

Figure 2.11 Typical bark of cherry (*Prunus* spp.).

leaves, whether the leaf margins are **serrate** or entire, and so on. One drawback to using identification keys in books is that the user can get stuck if unable to answer a question. Plant identification software usually does not have this drawback. Either way, an understanding of identification terminology is vital. This text provides some illustrations and descriptions of a few of these terms.

A few notes of caution about relying on keys are in order. First, a certain level of expertise and understanding of the terms used in the key is necessary to identify a plant correctly. Second, owing to seasonal differences in plant color and morphology, the plant being identified may not match its written characteristics exactly. Finally, genetic differences, plant condition and location, and the environment can affect the size of plant parts. Nevertheless, a key is a tool that can be helpful in determining a tree's identity.

Figure 2.12 Examples of conifer foliage types.

Chapter 2 Workbook

1. The classification of plants is called _______________.
2. List the levels of plant classification. The first letter of each term is given.

   K_______________

   P_______________

   C_______________

   O_______________

   F_______________

   G_______________

   S_______________ _______________
3. __________________ are vascular plants whose seeds are covered (by an ovary). __________________ are vascular plants with "naked seeds."
4. The term "dicotyledon (dicot)" refers to plants that have two seed leaves at germination. Grasses and palm trees belong to another group called _______________ and have only one seed leaf.
5. The naming of plants is called __________________.
6. Name five plant characteristics used to identify trees.

   a.

   b.

   c.

   d.

   e.
7. Draw a twig with the following leaf arrangements.

   opposite **alternate** **whorled**
8. Name a tree with palmately compound leaves: ____________. Name a tree with pinnately compound leaves: ____________ .
9. Draw a simple leaf with a lobed leaf margin.

10. Draw a compound leaf with serrate margins on the leaflets.

11. A compound leaf with multiple leaflets will have ______ bud(s).
12. Give an example of a tree that has more than one common name: ______________________.
13. In the scientific name *Acer saccharum*, *Acer* identifies the _____________, and *saccharum* identifies the ______________ ______________.
14. Species are often subdivided into ___________ or ___________ that have distinct differences from the general species.
15. A ______________ is a cultivated variety.

CHALLENGE QUESTIONS

1. Name three trees with characteristic scents that can be used in identification.
   a.
   b.
   c.
2. Draw a bipinnately compound leaf labeling the bud, petiole, a petiolule, and a leaflet.

3. Write and correctly label a scientific tree name with genus, species, variety, and cultivar.

SAMPLE TEST QUESTIONS

1. Douglas-fir (*Pseudotsuga menziesii*) differs from balsam fir (*Abies balsamea*) in that
   a. they are not in the same genus
   b. they are not in the same family
   c. Douglas-fir is actually a type of hemlock
   d. balsam fir is not a conifer

2. When two leaves and/or buds are located at the same node on a twig, the arrangement is called
   a. opposite
   b. alternate
   c. whorled
   d. compound

3. Select the scientific name that is written correctly
   a. *Quercus Rubra*
   b. *Quercus rubra*
   c. *quercus Rubra*
   d. *quercus rubra*

4. Which genus of trees usually does **not** have an opposite leaf arrangement?
   a. *Acer* (maples)
   b. *Fraxinus* (ashes)
   c. *Quercus* (oaks)
   d. *Cornus* (dogwoods)

5. Which conifers have needles in bundles?
   a. hemlocks
   b. firs
   c. pines
   d. spruces

Other Sources of Information

(See pages v–vi for complete bibliographic information.)

Dirr, 1998. *Manual of Woody Landscape Plants.*
Gilman, 1998. *Horticopia: Trees, Shrubs, and Groundcovers.*
Sunset Books, 1995. *Western Garden Guide.*

CHAPTER 3
TREE/SOIL RELATIONS

CHAPTER 1 TREE BIOLOGY

CHAPTER 2 TREE IDENTIFICATON

CHAPTER 3 TREE/SOIL RELATIONS

CHAPTER 4 WATER MANAGEMENT

CHAPTER 5 TREE NUTRITION AND FERTILIZATION

CHAPTER 6 TREE SELECTION

CHAPTER 7 INSTALLATION AND ESTABLISHMENT

CHAPTER 8 PRUNING

CHAPTER 9 TREE SUPPORT AND PROTECTION SYSTEMS

CHAPTER 10 DIAGNOSIS AND PLANT DISORDERS

CHAPTER 11 PLANT HEALTH CARE

CHAPTER 12 TREE ASSESSMENT AND RISK MANAGEMENT

CHAPTER 13 TREES AND CONSTRUCTION

CHAPTER 14 SAFETY

CHAPTER 15 CLIMBING AND WORKING IN TREES

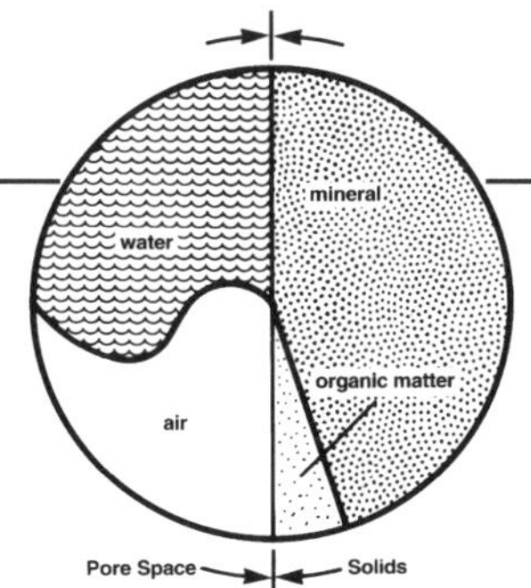

CHAPTER 3

TREE/SOIL RELATIONS

Objectives

1. Know the effects of soil texture and structure on soil compaction, drainage, water-holding capacity, and growth of tree roots.
2. Be able to define pH, cation exchange capacity (CEC), and buffering capacity. Describe their effects on the availability of essential elements to trees.
3. Understand the concept of the rhizosphere and the importance of soil organisms to tree roots. Know the function of mycorrhizae.
4. Understand the relationship between soil moisture, absorption of essential elements, and root growth. Define gravitational water, field capacity, permanent wilting point, and infiltration rate.

Key Terms

anion	compaction	leach	organic layer	soil profile
buffering capacity	field capacity	macropore	pH	soil structure
bulk density	gravitational water	micropore	parent material	soil texture
cation	horizon	mycorrhizae	permanent wilting point	symbiotic
cation exchange capacity	infiltration rate	nematode	rhizosphere	water-holding capacity
	ion	nutrient cycling		

INTRODUCTION

The relationship between tree root systems and the characteristics of the soils in which they grow has a greater influence on tree health than any other single factor. By knowing more about soil texture, structure, pH, and water-holding capacity, the arborist will be better equipped to manage trees in an urban environment.

PHYSICAL PROPERTIES OF SOIL

Native soils are the result of thousands of years of biological, chemical, and physical weathering and erosion of **parent material**, or underlying bedrock. Soils are usually dominated by the geology of the soil parent material. By volume, ideal soils are composed of 45 percent mineral materials (sand, silt, and clay), 50 percent open or pore space, and 5 percent organic matter and organisms (Figure 3.1). Over time, soils develop layers, called **horizons**, due to rainfall, heating and cooling, chemical reactions, and biological activities. The description and classification of these soil layers are part of a **soil profile** (Figure 3.2).

The soil profile normally consists of five major horizons (O, A, E, B, and C), although soil scientists classify a number of sublayers and transitional layers. Sometimes these layers can be distinguished by differences in color, which can indicate variations in drainage, organic-matter content, and other characteristic changes. The top of the profile, in an unaltered soil, is a thin layer of decomposing organic material called the **organic layer**. The next layer down is the A horizon. The A horizon contains most of the absorbing roots of trees. This horizon is normally rich in organic matter. The E horizon is an area of mineral weathering. The B horizon, intermediate

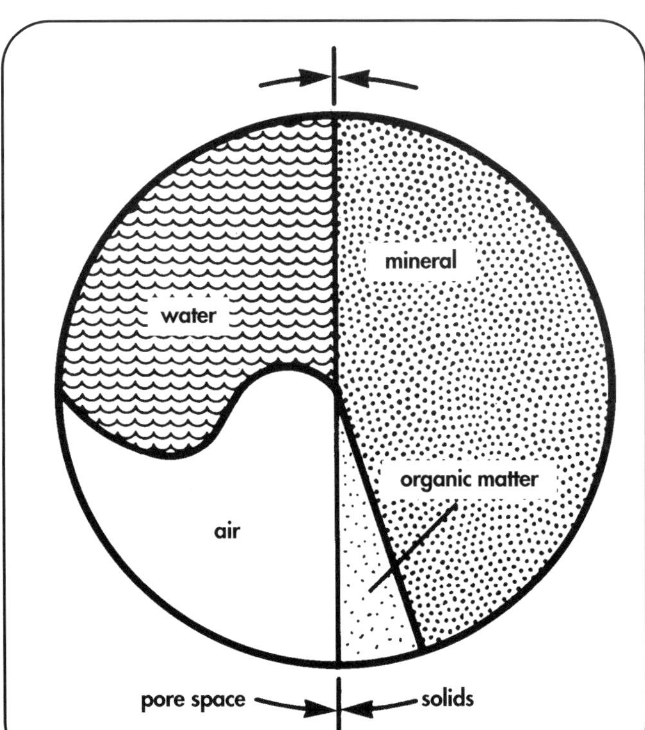

Figure 3.1 An ideal soil is 50 percent solids and 50 percent pore space, which contains water and air.

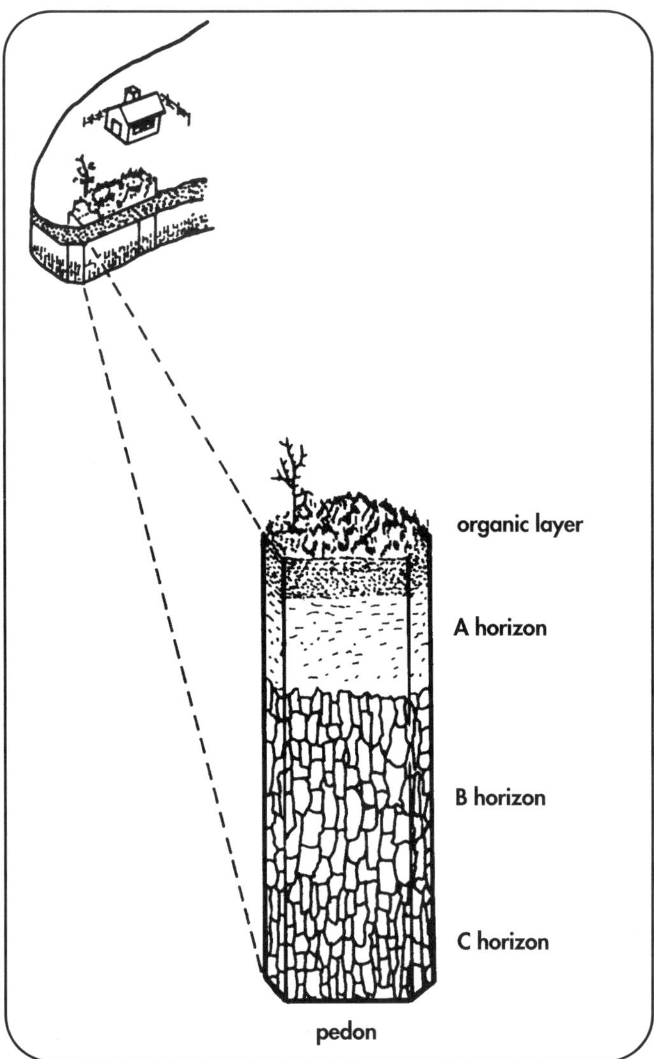

Figure 3.2 The soil profile shows the soil horizons.

in depth, is composed of fine-textured materials from the A horizon and soil particles from the lower parent material. The A and B horizons can be modified enough by the environment to be considered "topsoil." In urban situations, it is not unusual for much of the topsoil to have been stripped away or damaged during building construction. The C horizon, or subsoil, is the lowest layer other than the bedrock, and it is composed of the rocky parent material.

The organic layer, on the soil surface, contains such materials as leaves, twigs, bark, and organisms that compose the "litter" layer. The litter layer is very active biologically and is slowly broken up and decomposed by this activity. Minerals, decayed organic material, and other residues move down into the mineral soil. Soil is continually forming through the physical, chemical, and biological weathering of the parent material and the organic layer.

Tree roots grow where soil conditions are favorable. Roots require space among soil particles, organic materials and essential mineral elements, and adequate oxygen and water. Thus, absorbing roots are most frequently found in the upper 6 to 10 inches of soil, and tree roots are not usually found deeper than 3 to 4 feet.

Soil texture refers to the relative fineness or coarseness of soil mineral particles, specifically the proportions of sand, silt, and clay. Sand particles are relatively large, resulting in coarser-textured soils. Soils that have a high percentage of clay, the smallest soil particles, are fine textured. Textural classes of soil are determined by the percentage of the three particle types, as shown in the soil textural triangle. Loam is a mix of particle sizes often considered ideal because of the advantageous characteristics of its particle sizes. Soil texture affects a soil's ability to hold water and to provide oxygen to the roots. Thus, texture plays a role in determining which species of trees will do well in a given site (Figure 3.3).

Bulk density is the weight of dried soil per unit of undisturbed soil volume. Bulk density is a measure of a soil's porosity, or airspace between the particles. High bulk densities indicate a low percentage of total pore space. Large gaps between soil particles and aggregates are known as **macropores**. Macropores are normally filled with air. Small gaps between

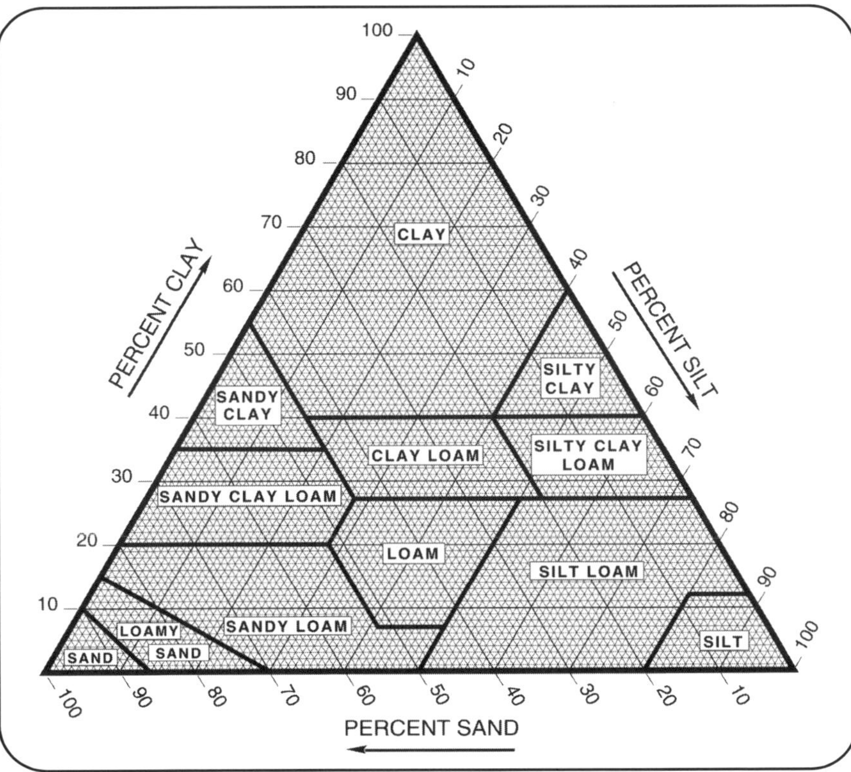

Figure 3.3 Soil texture triangle.

soil particles and aggregates are called **micropores** (Figure 3.4). They tend to be filled with water. Macropores dominate coarse soils (sand), yet fine soils tend to have more overall pore space, or lower bulk densities.

As a soil weathers, as organisms grow and die, and as organic materials are consumed and excreted, the basic soil particles become grouped or clumped together. These secondary groups or clumps are known as soil aggregates and provide **soil structure**. The size and shape of the soil aggregates, or particle clusters, are important in water and air uptake by tree roots. Soil structure modifies the influence of texture on pore space and drainage. For example, a well-developed granular structure facilitates aeration and water movement. Root growth, freezing and thawing, and burrowing insects and other animals contribute to changes in soil structure over time. These processes increase the ratio of macropores to micropores, improving soil aeration and root growth.

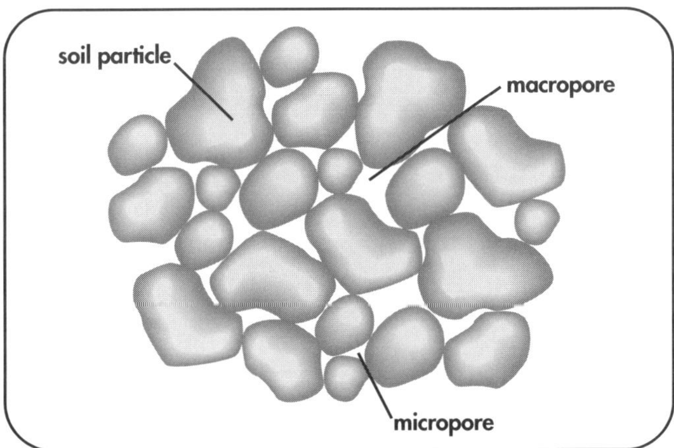

Figure 3.4 Pore space is the area between soil particles.

Soil aggregates, and accompanying pore spaces, are easily disrupted by physical compaction of the soil. When soil minerals are pressed together by vehicles, foot traffic, or other causes, pore space is greatly reduced, especially macropores that tend to be oxygen filled. Soils that have a broad range of particle sizes, and those that are wet, can be severely compacted. When a soil is severely compacted, the soil particles at the surface often align parallel to the surface, much like shingles on a roof, inhibiting water percolation. Compacted soil restricts root growth, reduces water infiltration and availability, and limits the movement of oxygen and carbon dioxide in the root zone.

CHEMICAL PROPERTIES OF SOIL

Soil **pH** is a measure of the acidity or alkalinity of soil. The pH scale ranges from 0 to 14. A pH of 7 is considered neutral. Soils with a pH less than 7 are acidic, and those with a pH greater than 7 are alkaline. Because pH is a logarithmic function, a pH of 6 is ten times more acidic than a pH of 7, and a pH of 5 is 100 times more acidic than a pH of 7. Although variable for different tree species, a pH in the range of 6.0 to 6.5 is generally favorable for most plants.

Soil pH has many effects on the ecology and chemistry of the soil. The pH may affect which species will grow and which soil organisms are present. One important effect of pH on tree growth is the availability of minerals. At certain pH levels, essential elements form chemical compounds that are insoluble in water and unavailable to plants because roots can only take up minerals dissolved in water. For example, in highly acidic soils with a pH of 5.5 or below, phosphorus may be deficient while other elements may become toxic. In alkaline soils, iron and manganese may be unavailable because the chemical form of these elements changes, and they become solid particles. However, the availability of calcium, magnesium, and potassium may increase with higher pH.

It is difficult to alter soil pH to achieve a more desirable growing medium. Sulfur may be added to lower the pH, or lime may be added to raise the pH. However, altering the pH is usually not practical for arborists because the volume of soil in the tree root area is so great. In addition, many soils have a high **buffering capacity**, or resistance to change in pH, especially soils high in clay or organic matter.

Minerals required for tree growth (essential elements) dissolve in water, making them available for absorption by tree roots. In solution, these elements are charged particles called **ions**. Negatively charged ions are called **anions**; positively charged particles are called **cations**. The **cation exchange capacity** (CEC) is a measure of the soil's attraction, retention, and exchange of positively charged cations. Organic matter and clay normally carry a negative charge. This negative charge attracts and holds cations, giving soils high in clay and organic matter a high cation exchange capacity (Figure 3.5). This makes them more fertile than coarse-textured soils. The attraction between cations and soil particles minimizes the tendency of minerals to **leach**, or wash through the soil.

BIOLOGICAL PROPERTIES OF SOIL

Soil is an ecosystem containing billions of organisms. Although some soil organisms may damage roots, many are beneficial, and others have no direct effect on tree roots. Animals such as insects and roundworms, which inhabit the soil and litter layer, increase aeration and accelerate decay. Other animals feed on roots. **Nematodes**, microscopic roundworms, can parasitize tree roots, and some transmit disease. Other nematodes feed on pathogenic, disease-causing organisms. Other organisms found in the soil ecosystem are bacteria and fungi. Most of these are an essential part of the balance of life, helping to decompose organic matter. A few cause diseases in plants.

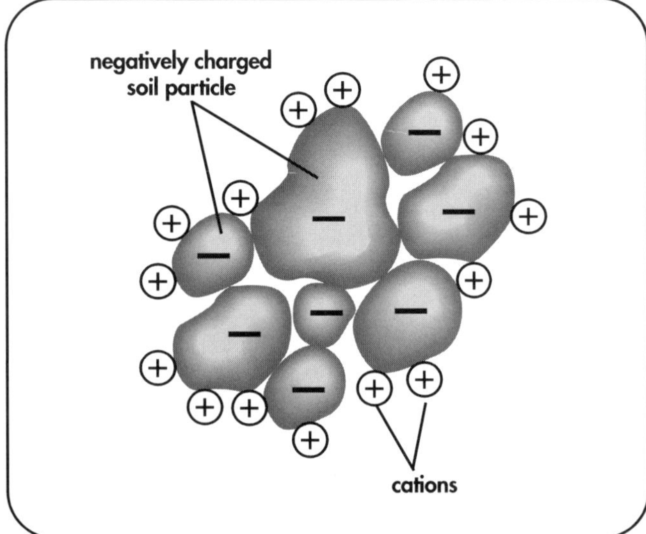

Figure 3.5 Cations are attracted to and held by negatively charged soil particles.

The **rhizosphere** is the zone of intense biological activity near the actively elongating roots. As roots extend through the soil, the root caps and external layers are sloughed off and materials from the roots are released into the soil. This is a constant source of organic matter on which microorganisms feed. The rhizosphere is an altered environment within the soil where many creatures flourish.

Mycorrhizae, literally meaning "fungus roots," are special, fungi-infected roots of most plants. Mycorrhizal fungi live in a **symbiotic** relationship with the roots. This means that the fungi and roots both benefit from the relationship. The roots provide a place for the fungi to live and provide food. The fungi increase the roots' ability to absorb water and essential elements, especially phosphorus (Figure 3.6).

Figure 3.6 Mycorrhizae aid in the uptake of water and minerals.

Nutrient cycling is especially important in natural plant systems. As a plant grows, roots absorb essential mineral elements from the soil solution and produce new woody material and leaves. As seasons pass, plants or plant parts die and are returned to the soil surface where they are acted upon by soil organisms and the weathering processes. Gradually, decomposition occurs and nutrients are released into the soil where they become available, once again, to plant roots.

SOIL MOISTURE AND PLANT GROWTH

The amount and size of soil pores and the total surface area of the particles determine the amount of water that a soil can hold. Clay soils have more total pore space and particle surface area than sandy soils. Thus, clay soils have a higher **water-holding capacity** than sandy soils. Water that drains from the larger macropores under the force of gravity is called **gravitational water**. A soil is said to be at **field capacity** when gravitational water has drained away. Water that remains is held by the soil particles (Figures 3.7 and 3.8).

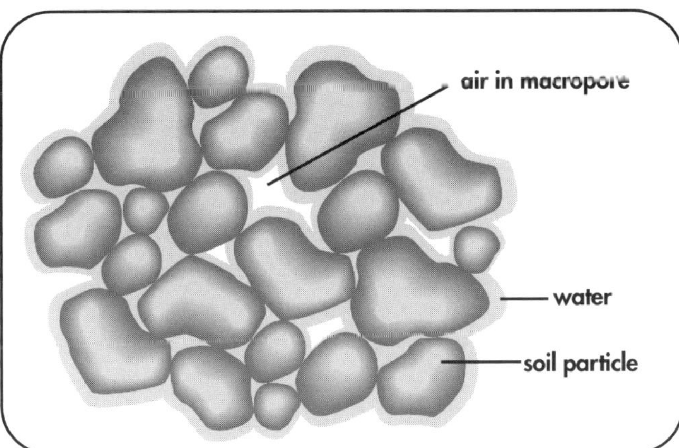

Figure 3.7 Macropores tend to be air filled at field capacity, but water remains in the micropores, adhering to soil particles.

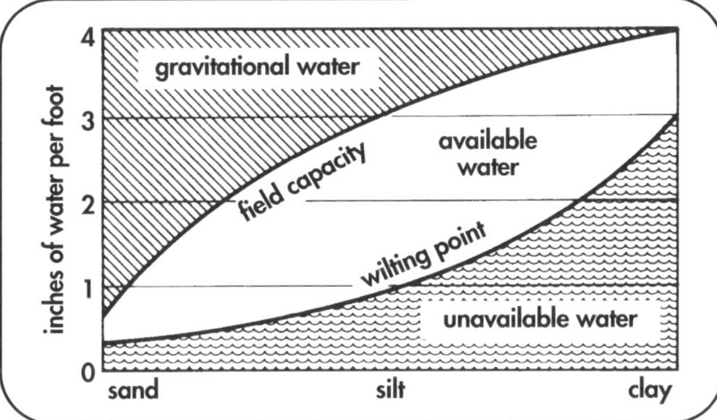

Figure 3.8 Soil texture influences the amount of water available to a tree.

Once a soil is at field capacity, water is absorbed by plant roots or it evaporates. Roots draw water from the soil as long as they can overcome the forces of adhesion, which hold the water to soil particles. Tree leaves may wilt during periods of high water demand during the day but usually recover at night when transpiration decreases. Eventually, depending on the water-holding capacity of the soil, a point will be reached at which a tree cannot pull any more water from the soil. This point is called the **permanent wilting point**. A plant will not recover from wilting unless water is added to the soil. Trees existing in soil at the permanent wilting point, when no usable water is available from the soil, usually decline and die.

Trees need both air and water, and there is a delicate balance of water and gases (mostly oxygen, nitrogen, and carbon dioxide) in the soil pores. Tree roots require oxygen, but they also give off carbon dioxide as respiration takes place. Gas exchange between the soil and the atmosphere generally occurs by diffusion through the soil surface. Oxygen levels tend to be higher near the surface. If there is insufficient gas exchange, such as in saturated or compacted soils, carbon dioxide levels may build up and there may be an oxygen deficit. This can reduce root function and growth, and if it occurs for prolonged periods, can lead to root death.

Layers of varying soil texture are common in urban soils and can be a problem in planting situations. The texture of the soil plays a major role in the **infiltration rate** of water. If a layer of coarse soil (sand) is on top of a fine-textured soil (clay), water will accumulate in the upper layer as it slowly infiltrates into the lower layer. Inversely, if the coarse layer is below the fine layer, the water will not drain into the lower layer until the upper layer is completely saturated (Figure 3.9).

If a tree is planted in clay soil and the backfill is amended with a more coarse soil, the planting hole may act as a bowl and retain the water (the "teacup effect"). If the water is held for a long time, the roots drown. Placing gravel in the bottom of planting holes, pits, or containers does not improve drainage because the water remains in the finer-textured soil until saturation is reached.

URBAN SOILS

While natural conditions create most forest soils, human activity is the principal influence on urban soil, often degrading the soil's natural characteristics that benefit trees. Urban soils often do not have an organic layer. The soil may be compacted or crusted and may have a disrupted soil profile, altered drainage, elevated pH, or subsurface barriers as a result of building foundations, roads, or underground utilities. All of these factors may impact root growth and tree health, eventually leading to decline.

Turf, bare ground, or hardscape replaces the organic layer in many urban soils. Hardscape may impair aeration and water infiltration. Organic-matter reduction decreases biological activity, hampers soil structure development, and interrupts nutrient cycling.

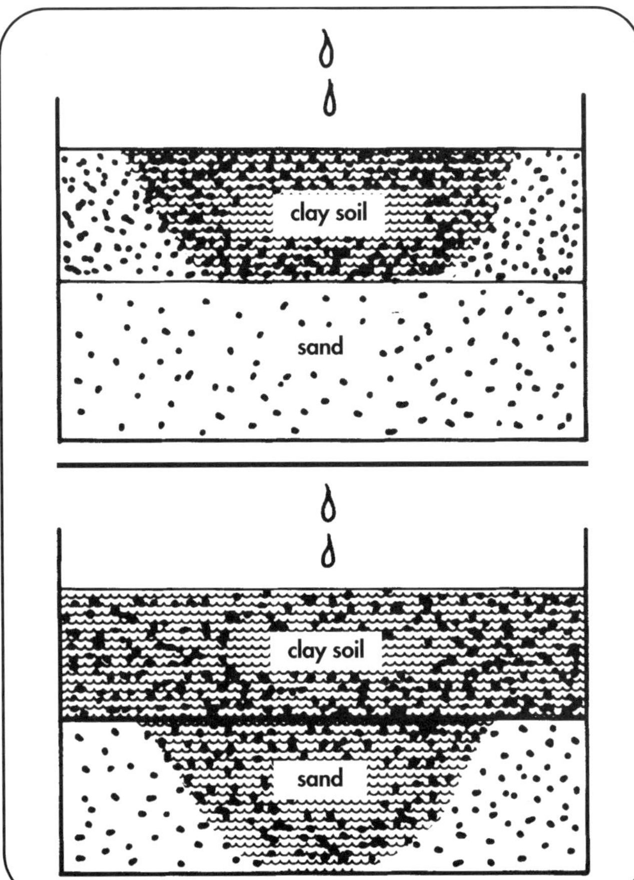

Figure 3.9 This diagram illustrates water infiltration through a layer of fine-textured soil into sand. The upper layer becomes completely saturated before water drains into the lower layer.

Urban soils may lack important microorganisms such as mycorrhizae. Furthermore, the absence of the insulating forest organic layer and vegetation can contribute to greater fluctuations in temperature extremes.

Compaction is one of the biggest problems in urban soils. Compaction is often caused by construction, foot or vehicular traffic, engineered soils to support roads or buildings, or other factors. Compaction reduces total pore space and the proportion of macropores to micropores. Loams, silt loams, and other soils with a variety of particle sizes may be particularly vulnerable to compaction because small particles are pressed into large pore spaces between coarse particles.

Trees and soils are so ecologically interdependent that it is hard to imagine separating them from one another. Yet the processes involved with urban development disrupt this ecological balance, creating growing conditions that may range from unfavorable to antagonistic. It has been said that the vast majority of tree decline situations can be attributed to an initial soil stress. Trees are living systems driven by energy. They must obtain sufficient oxygen, water, essential elements, and other components from the soil to meet their energy requirements. Understanding soil is vital to arboriculture because soil is, quite literally, the foundation within which a tree grows.

Chapter 3 Workbook

1. The majority of the fine, absorbing roots of a tree are in the ___ and___ horizons.
2. Driving vehicles across wet soil will _______________ the soil and destroy soil ________ _____________.
3. If lime is added to a soil to change the pH, but there is no significant change, which soil property is preventing pH change? __________________________ __________________________
4. True/False—Negatively charged clay particles hold cations near their surface.
5. In 1 foot of clay soil with 4 inches of water in the soil, approximately how many inches of water would be available to the tree? ______________.

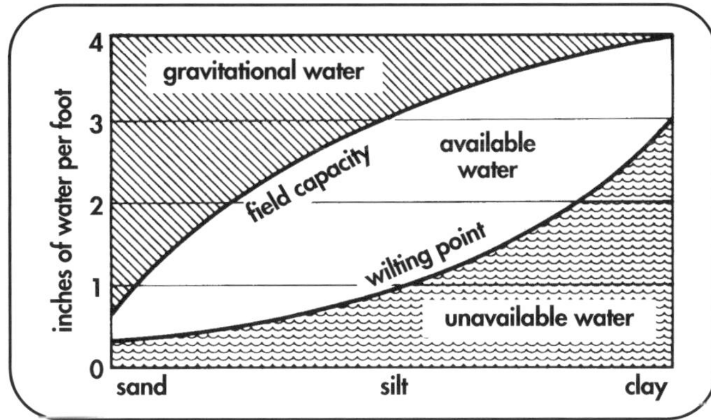

6. _______ ________________ modifies the effects of soil texture as the soil particles form clumps or aggregates.
7. True/False—Soil can hold water so tightly that the ability of tree roots to absorb the water is restricted.
8. Soil texture refers to the relative coarseness or fineness of a soil. Rank the following from the most fine texture (1) to the most coarse (3).

   _____ silt _____ clay _____ sand
9. On the pH scale, less than 7 is __________, 7 is ___________, and more than 7 is ____________.
10. True/False—Soil structure can be more important than soil texture in determining oxygen and water availability.
11. A pH of 5 is ________ times more acidic than a pH of 7.
12. The process in which dissolved minerals wash down through the soil profile and are lost is called _______________.
13. Many essential elements are dissolved in soil water in the form of positively charged particles called ____________.
14. If the soil is too ____________, iron and manganese may be in a chemical form that is unavailable to trees.
15. The buffering capacity is the resistance of a soil to changes in pH. Clay soils and soils high in organic matter usually have a _______ buffering capacity.
16. The ________________ is the zone of intense biological activity near the actively elongating roots.
17. True/False—All of the water in the upper 12 inches of soil is available to the tree roots.

18. Water that drains from the macropores is called ____________________ water. Following drainage, the soil is said to be at ________ ________________.

19. True/False—Most soil organisms cause disease or decay in tree roots.

20. True/False—Many tree roots exist in a symbiotic relationship with fungi that assist the tree in water and mineral absorption.

MATCHING

_____ sand	A. "fungus roots"
_____ buffering capacity	B. measure of acidity/alkalinity
_____ field capacity	C. fine-textured soil particles
_____ rhizosphere	D. water that drains from the macropores
_____ macropores	E. ability of a soil to attract and absorb cations
_____ mycorrhizae	F. coarse-textured soil
_____ CEC	G. ability to maintain pH
_____ clay	H. soil after gravitational water has drained
_____ pH	I. soil zone around roots
_____ micropores	J. tend to be air filled
_____ gravitational water	K. tend to be water filled

CHALLENGE QUESTIONS

1. How do soil texture and organic matter influence the buffering capacity of a soil?

2. Why are arid soils generally alkaline and high in salts?

3. Why is hygroscopic water, the water remaining in the soil at permanent wilting point, unavailable to plant roots?

4. Explain why a planting container filled completely with soil drains better than a container with gravel in the bottom.

5. If, during construction, most of the A horizon and organic layer is stripped away, leaving the B horizon as the uppermost soil, what would be the effects on newly planted trees in the landscape?

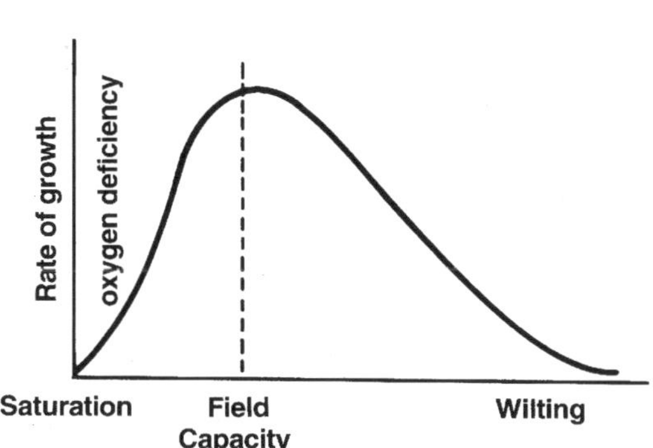

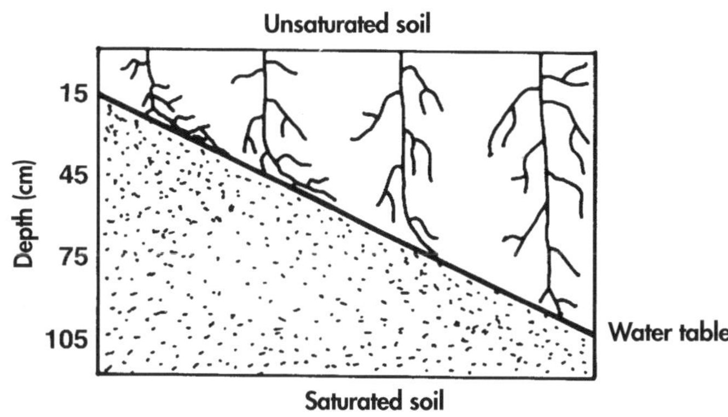

6. Based upon the graphs above, explain how root growth is affected by excess water in the soil, especially with a tree adapted to an upland site.

7. How much water would you need to add to a 5-foot-deep clay soil at permanent wilting point to bring it up to field capacity, if its field capacity is 4 inches per foot and its permanent wilting point is 2 inches per foot?

SAMPLE TEST QUESTIONS

1. The primary factor in controlling the depth, spread, and distribution of tree roots in the soil is/are the
   a. available oxygen and water
   b. genetics of the plant
   c. pH of the soil
   d. available phosphorus and potassium

2. The most fibrous and absorptive portion of a tree's root system is found
   a. in the upper 6 inches of soil
   b. deeper than 12 inches in the soil to protect against drought and temperature extremes
   c. immediately surrounding the trunk
   d. connected to the taproot

3. If a planting hole in a clay soil site is backfilled with sandy soil,
   a. drainage will be improved, helping the tree to establish
   b. nutrients will be more available to the newly established roots within the planting hole
   c. water will drain very slowly out of the planting hole
   d. the improved texture of the backfill will reduce the chances of girdling roots forming later

4. When soil is compacted,
   a. micropores combine to form macropores
   b. soil aggregates are broken up, giving the soil a finer texture
   c. a high water content will reduce the damaging effects
   d. total pore space and the percentage of macropores are reduced

5. A characteristic of sandy soils in arid regions is
   a. they tend to become alkaline, and salts build up due to the lack of heavy rainfall
   b. they tend to become acidic because basic ions leach out
   c. they are fine in texture due to the high sand content
   d. they have a high water-holding capacity because rainfall is scarce

Other Sources of Information

(See pages v–vi for complete bibliographic information.)

Craul, 1999. *Urban Soils: Applications and Practices.*
Harris et al., 1999. *Arboriculture: Integrated Management of Landscape Trees, Shrubs, and Vines.*
Shigo, 1986. *A New Tree Biology.*

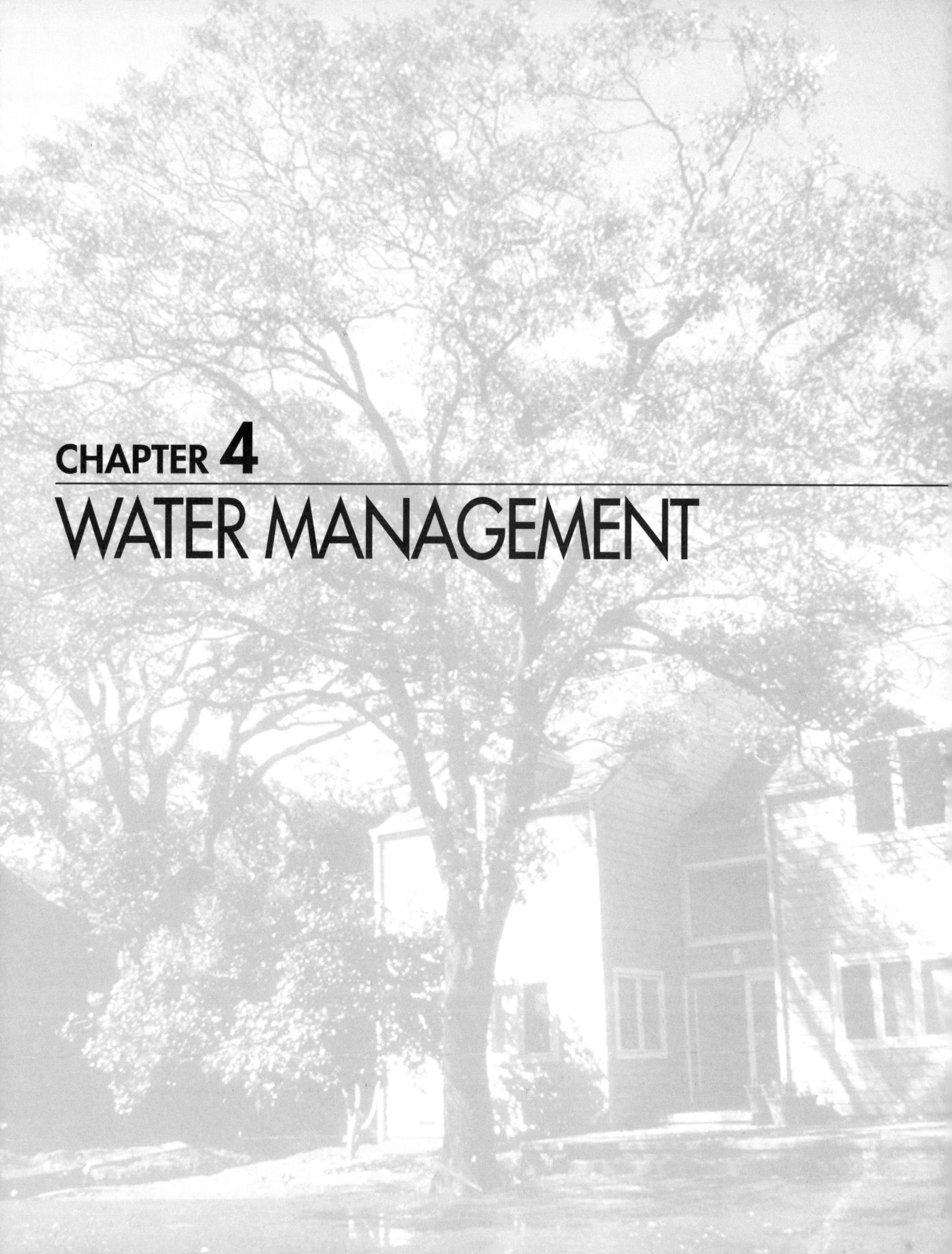

CHAPTER 4
WATER MANAGEMENT

CHAPTER 4 WATER MANAGEMENT

CHAPTER 4

WATER MANAGEMENT

Objectives

1. Describe water availability and the movement of water in the soil.
2. Understand the effects of excess or insufficient soil water.
3. Explain why irrigation is important in many urban landscape sites.
4. Become familiar with the different methods of irrigation and with the advantages and disadvantages of each. Understand what is meant by "minimum irrigation" and when it is used.
5. Understand the principles of evapotranspiration, infiltration, and water-holding capacity.
6. Explain why good drainage in the root zone of trees is important.
7. Discuss the advantages of mulching around trees. Know the various materials that are commonly used for mulch and how they are properly applied.

Key Terms

antitranspirant	evapotranspiration (ET)	minimum irrigation	tensiometer
available water	infiltration	percolation	turgid
desiccation	infiltration rate	phytotoxic	water-holding capacity
drip irrigation			

INTRODUCTION

Water is vital to plants. Trees absorb water and minerals (dissolved in water) from the soil. A large tree can absorb hundreds of gallons of water in a day, and as much as 95 percent of that water may be transpired into the air. Over time, without sufficient soil moisture, a tree cannot take up essential elements, photosynthesis is reduced, and the tree will decline. Yet too much water in the root zone also can damage trees.

Irrigation is a necessity in many urban planting sites. Water can be applied in the amounts desired and at suitable times with irrigation systems. However, if drainage is not adequate, if the tree is not irrigated wisely, or if the irrigation water source has a high soluble-salts content, root problems or high soil salinity (salt buildup) can occur.

Good water management needs to consider the water requirements of the plants, soil relations, irrigation frequency and timing, water conservation, application techniques, and drainage.

PLANT SOIL AND WATER REQUIREMENTS

The amount of water needed by plants varies with the species and size of the plant, the air temperature, humidity, light levels, and wind movement over the leaves. Transpirational water loss from leaves is controlled by stomatal opening and closing in response to the environment. When transpirational water losses exceed the plant's ability to take up water from the soil, the plant wilts. In an extended period of water deficit (drought), plants respond by wilting, dropping leaves, developing modified leaves, or increasing the production of absorbing roots. Severe drought periods can cause extensive root loss, leaf abscission, and plant death.

The **available water** is the amount of water in the soil between field capacity and permanent wilting point. The amount of water in the soil that is available to plants differs with soil textures. Clay soils have a greater **water-holding capacity** than sandy soils. As a result, sandy soils need to be irrigated more frequently. Soil characteristics also affect the way water moves in the soil. **Infiltration** is the term for water movement into the soil; **percolation** describes water movement within the soil. Water moves downward faster and more efficiently than it moves sideways, although this movement, too, is affected by soil texture. Because clay soils have a much slower **infiltration rate** than sandy soils, water needs to be applied slowly over long periods of time.

Above field capacity, water moves downward through the soil in response to gravity as excess water drains from the macropores. Below field capacity, water moves through micropores from areas of higher moisture content to areas of lower moisture content (Figure 4.1). This movement can be in any direction, including upward from the water table. Compaction and changes in soil texture layers can interfere with normal percolation.

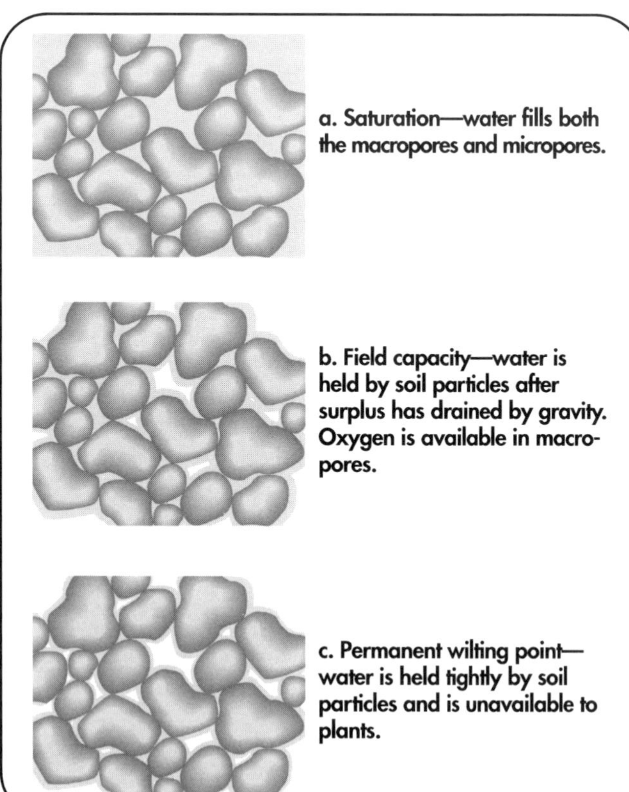

Figure 4.1 Soil moisture conditions.

Trees and other landscape plants vary by species in their normal patterns of water use and requirements. Most species can be categorized as "water spenders," using relatively large amounts of water. They tend to develop large, spreading root systems that allow them to make use of soil moisture over a great area. They cannot tolerate long drought periods without water. Examples include elm *(Ulmus)* and eucalyptus *(Eucalyptus)*.

Some species are more drought tolerant and are adapted to lower rates of water use. Of these, some adapt by becoming virtually dormant during extended dry periods. Others, "water conservers," reduce water loss through their foliage. They tend to have small, thick, leathery leaves, sometimes with sunken stomata. Examples include olive *(Olea europea)* and mesquite *(Prosopsis glandulosa)*. Drought-tolerant plants not only are adapted to survive on minimal water, they can be adversely affected by supplemental irrigation water.

DROUGHT

Plants differ in their water requirements. Species that are adapted to long, dry summers can tolerate four or five months without rain or irrigation. These species tend to be small and tend to grow farther apart naturally. They tend to have deeper, more extensive root systems. Most also have foliage adapted to reduced transpiration, with smaller, thicker leaves with fewer stomates.

Trees that are accustomed to fairly regular rainfall throughout the summer may begin to show stress after several weeks without rain. An early symptom is wilting of the foliage. Wilting is common during hot, summer afternoons, but if the leaves are not **turgid** (fully hydrated) again in the morning, the tree may be seriously stressed. Leaves may begin to turn brown or drop. If water is not added, roots will be lost and the tree will decline (Figure 4.2).

Drought following periods of moderate to high soil moisture conditions can be particularly damaging. Moist conditions encourage shallow rooting. If the tree does not produce some roots deeper in the soil, it may not be able to take up enough moisture when the soil near the surface becomes very dry.

FLOODING

Flooding can be very damaging to trees. Lack of oxygen in the soil suffocates roots, changes the chemical composition of essential elements, can create mineral

Figure 4.2 Trees not adapted to "heat islands" such as parking lots are particularly prone to water stress injury. This zelkova (*Zelkova*) was seriously injured when water was not provided, while similar trees in lawn areas were in good condition.

toxicities, and can lead to fermentation in root cells. The longer the excess water remains, the more tree health deteriorates. After just a few hours under flooded conditions, photosynthesis stops, transpiration slows, and some soil organisms are killed. After several days, some root loss is experienced in most species. If conditions persist, some species will decline and die.

For species that survive flooding, there can still be long-term consequences. Some will be predisposed to stress factors including drought and pests, and they may not be able to generate an adequate defense. Trees that have experienced flood conditions are prone to toppling due to root loss and wet soils, and they suffer a higher incidence of root or collar rot. In addition, grade changes due to erosion and soil deposition can lead to long-term stress.

Trees that survive flood conditions should be inspected for hazards and structural integrity of the root system. Changes in grade should be corrected if possible, and measures to improve soil aeration should be considered. Soils should be analyzed to determine whether mineral deficiencies exist before consideration is given to fertilization.

IRRIGATION

The most basic principle of watering is that enough water should be supplied to the soil to replace what the plant uses, and what is lost to evaporation and percolation (Figure 4.3). Water requirements of trees vary by species and age, as well as by numerous environmental factors. Newly transplanted trees of any species require frequent water addition, especially within the root ball. The same frequency of irrigation on many mature specimens can lead to problems, however.

As a rule, frequent, shallow waterings encourage surface rooting—which makes the tree more vulnerable to **desiccation** (drying out) during periods of drought. Infrequent, deep soakings encourage the production of a deeper root system and more

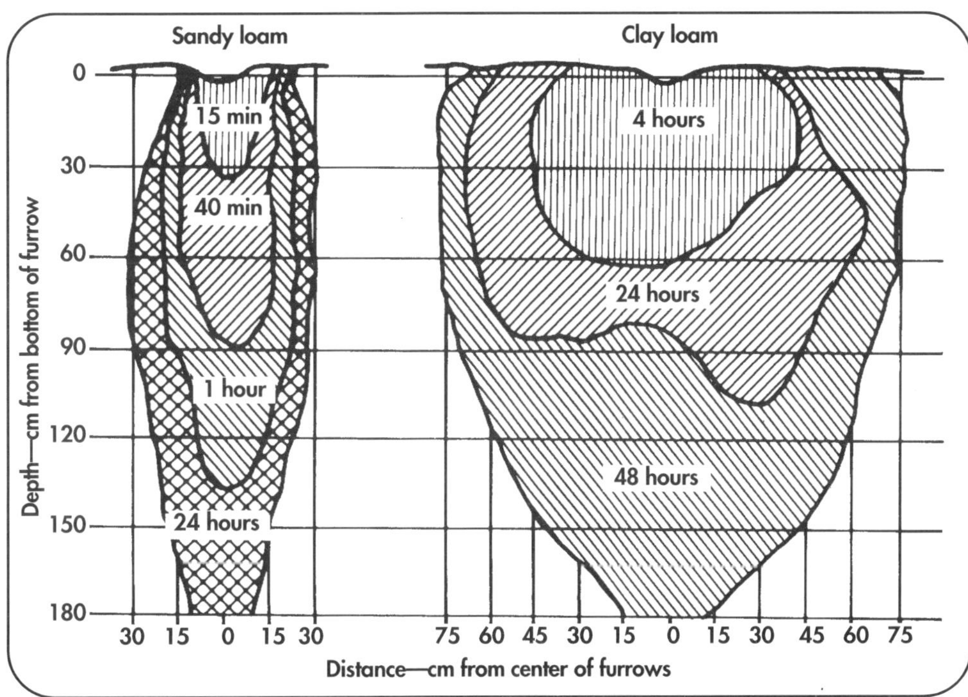

Figure 4.3 A light irrigation wets soil to field capacity at a lesser depth than a heavy irrigation. Note the difference in wetting patterns for sandy and clay loam soils.

drought-tolerant trees. If the soil is allowed to dry between irrigations, natural shrinking and swelling improves soil structure. Conversely, frequent, shallow irrigation tends to compact the soil surface and reduce the rate of water infiltration.

The most beneficial time to irrigate plants is during the late night and early morning hours. Evaporation is minimized, and the foliage has time to dry thoroughly during daylight hours. Evening watering is efficient for water use, but should be applied after dew is on the leaf surfaces. Where fungal diseases of the foliage are a problem, it is important not to extend the period in which the leaf surface is naturally wet from dew.

Water should be distributed evenly to as much of the root system as possible. Watering the lower trunk (root collar) should be avoided because it can lead to increased fungal decay problems for the tree. Topography (hills and valleys) affects water distribution. Soil tends to dry faster on hills, while water may accumulate in valleys and low areas. The water application rate should not exceed the soil infiltration rate.

Figure 4.4 Trees can be irrigated in a variety of ways. Here sprinklers are used to supply water to the extensive root system of this oak (*Quercus*). It is important to avoid irrigation around the root crown, however.

If water is applied too quickly, runoff can cause erosion problems and reduced infiltration. Puddling or runoff that results from high application rates or poor irrigation design wastes water and can be detrimental to root growth and function.

Sprinkler irrigation is currently the most common method of watering landscape plants. Sprinklers can apply water uniformly throughout a planted area. In most sites, it is best to choose sprinkler heads with low application rates and those that minimize the amount of water on the foliage. This method will reduce runoff and fungus problems (Figure 4.4)

One disadvantage of sprinkler irrigation is surface compaction created by water hitting the soil surface and dispersing the soil aggregates. This situation can be especially troublesome when sprinklers apply too much volume to be absorbed by the soil, causing significant runoff. Soil crusting and erosion from sprinkler irrigation can be minimized with the use of mulch. With sprinkler irrigation, less frequent waterings of longer duration will lessen the likelihood of salt buildup. If irrigation water does not penetrate the soil deeply, most of the salt brought in by that water will remain near the surface. Under saline conditions, irrigation may need to be provided more frequently. Soil damage can result from sodium in irrigation water (Figure 4.5)

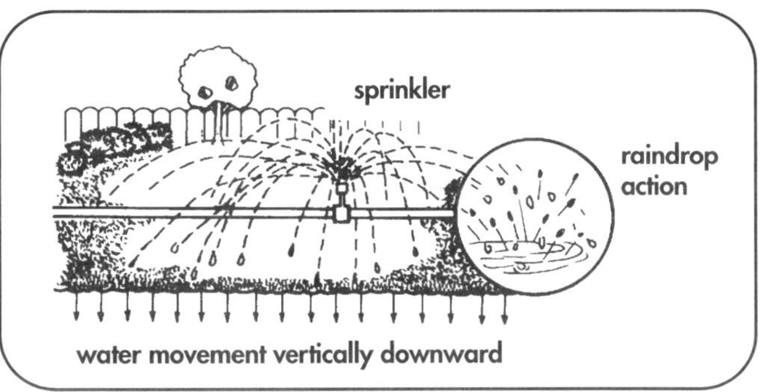

Figure 4.5 Overhead irrigation. Note that splashing water can compact and erode soil, and water can accumulate if the application rate is high.

Drip irrigation is used to conserve water, allowing more water to be absorbed with less loss by evaporation. Water is applied very slowly over a long period of time. Drip emitters should be spread out over the entire root zone and kept away from tree trunks. They should be moved out farther from the trunk each year until the plant is established, in order to apply water to the expanding root system. Emitters can easily become clogged, and the system should be checked regularly for proper operation.

Other methods employed to irrigate trees include high-pressure water injection (Figure 4.6), soaker hoses, basin irrigation, and temporary, portable drip systems used for establishment of young trees or during droughts. These approaches are usually used for single trees or small landscaped areas. High-pressure water injection can deliver water deep in the root zone but may bypass most of the absorbing roots at the surface. Soaker hoses are portable and can apply water slowly where runoff may be a concern. Basin irrigation is useful for newly planted trees. It provides even water distribution for a slow soaking but should be avoided where drainage is a problem. Portable systems vary in type and size but can be an effective means of supplying water to the root balls of newly planted trees.

Figure 4.6 Pressurized water injection is sometimes used to irrigate trees growing in compacted soils or where surface barriers (concrete, asphalt, etc.) exist.

Most irrigation systems are installed primarily to irrigate turf and other landscape plantings. Too often, trees are not properly considered in application techniques and rates. Excessive irrigation of trees is becoming a problem in many parts of the country. Fungal problems, especially root rot and collar rot, are becoming more prevalent. These conditions are particularly troublesome in instances where irrigation is applied directly on the trunks of trees. Tree species that are adapted to dry summers and native to such climates are prone to problems associated with frequent irrigation.

Minimum Irrigation

With water at a premium in some areas of the country, the principle of **minimum irrigation** is important to consider. Minimum irrigation is designed to maintain plants during periods of reduced rainfall. In some cases, this may be as little as one or two supplemental waterings in a growing season. The amount of water needed to maintain acceptable quality (characteristic canopy density and leaf color) for trees differs from species to species.

When planning a new landscape where minimum irrigation is to be used, plants with similar water requirements should be grouped together and shade used wisely. This technique usually employs the use of plants that require little water and highly efficient irrigation methods. Keep in mind that new plantings require more frequent waterings until they develop established root systems. Also, caution is in order when transitioning to minimum irrigation because plants must adapt to lower soil moisture conditions. Minimum irrigation practices require an understanding of soil–plant–water relations: plant and soil water loss (**evapotranspiration**, or **ET**) (Figure 4.7), water-holding capacity, application rate, infiltration rate, and irrigation system efficiency. Evapotranspiration is measured in many states and is used to estimate water loss from agricultural crops and turf grasses. Recent attempts have been made to adapt this information for trees and other landscape plants. Charts of average ET are often available through Cooperative Extension or Agricultural Department offices. Minimum irrigation combines water loss information with the amount of water available in the soil to determine appropriate irrigation schedules.

Another approach to minimum irrigation employs soil moisture monitoring. Soil moisture sensors, **tensiometers**, are used to measure soil wetness or dryness. Carefully tracking soil moisture before and after irrigation allows the amount of available water to be assessed in order to adjust irrigation recommendations.

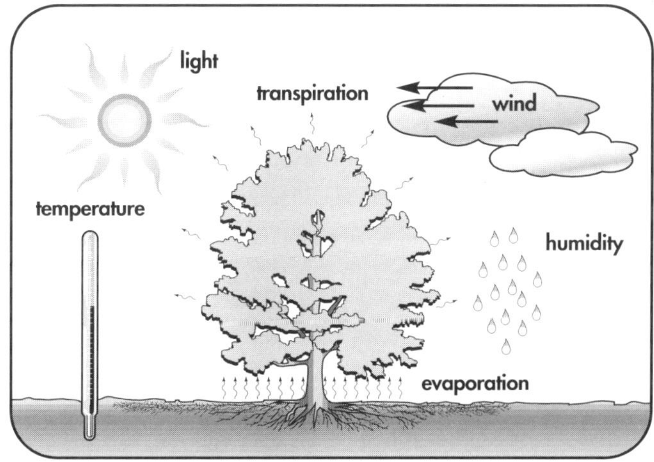

Figure 4.7 Evapotranspiration rates are dependent on environmental conditions, including light, temperature, wind, and humidity.

Arborists and other professionals may be called upon for advice when irrigation systems are being designed, installed, or adjusted. An understanding of the water requirements of trees and other plants in the landscape is an essential part of estimating how much water will be needed and at what frequency. Information regarding ET, water-holding capacity, irrigation efficiency, infiltration, and application rates needs to be put together so that an appropriate watering schedule can be developed.

Irrigating with Recycled Water

The water source for irrigation affects the quality of the water. As water flows over or through soil and rock, mineral salts dissolve and are picked up. Excess salts in the water can be detrimental or even toxic to plants. Recycled water can be high in salts as well as other minerals. These minerals can raise the soil pH, increase soil salinity, cause problems with phytotoxicity, and clog irrigation nozzles. Recycled water can also be high in essential elements such as nitrogen, phosphorus, and sulfur.

The success rate in using recycled water depends mostly on the soil type, water quality, and sensitivity of the plants to salts. Because good drainage is essential, sandy soils usually work better. More frequent irrigation is usually necessary because drought stress will occur at a higher soil moisture content under saline conditions. It is important to monitor soil conditions and plant health frequently when irrigating with recycled water.

WATER CONSERVATION

Conservation of water is becoming a bigger concern as the population grows, especially in more arid climates. Landscape managers and arborists must be more aware of plant water use and must keep water efficiency in mind when planting and maintaining trees. In addition to the principles of minimum irrigation already discussed, there are a number of ways to reduce water use.

Mulches

Mulches are materials placed on the soil surface to reduce moisture evaporation and improve soil conditions. Mulching the soil surface has several benefits. A layer of mulch reduces water evaporation from the soil. It minimizes weed competition (and thus water use), reduces soil erosion, and can improve soil aeration. The soil will be kept cooler in the summer and warmer in the winter. Most landscapers and home owners think that a mulched planting bed has a neat, finished appearance. Trees that are mulched are less likely to fall victim to lawn mower damage and will develop a more extensive root system.

A wide variety of materials are used as mulch. Organic mulches include wood chips, bark, pine needles, or other plant waste materials. Inorganic mulches such as crushed stone are also used. The advantage of organic mulches is that they add organic matter to the soil as they decompose, enhancing soil fertility and aeration. Some organic mulches need to be replenished every year or two. If stone is used, the type should be taken into consideration. Crushed limestone, for example, can raise the soil pH. Inorganic mulches provide fewer benefits to the tree.

In most landscapes, the mulch layer should not exceed 2 to 4 inches in depth and should not be placed against the trunk of the tree. Collar rot can occur in trees when mulch or soil is held next to the aboveground bark of the trunk. Geotextile fabrics are sometimes placed under the mulch and can reduce weed problems. However, a layer of plastic under the mulch should not be used because it will reduce water and air infiltration in the root zone.

A number of methods have been employed to attempt to increase the water-holding capacity of soils, especially sandy soils or container soils. Incorporating organic matter can be beneficial, but the effects are short-lived because the organic matter decomposes. Research continues on other soil additives for increasing water-holding capacity, but none has gained widespread use in arboricultural practices.

Antitranspirants

Antitranspirants are chemicals sprayed on plants to reduce water loss through transpiration. The spray forms a thin coating on the foliage, which reduces water loss through the stomates. Antitranspirants have been used to increase the chances of transplant survival during stress periods. They have also been used to help get plants through drought conditions. Antitranspirants are also used to prevent desiccation (drying out) of broadleaf evergreens during winter months, particularly when the ground is frozen.

Caution should be practiced when antitranspirants are used. Studies have shown them to be **phytotoxic** (poisonous) to some species. The effectiveness of antitranspirant sprays in reducing water loss depends on the species of tree and environmental conditions such as temperature and humidity. Some antitranspirants crack and peel in response to cold temperatures and the drying winds of winter. If antitranspirants are used in full sun, heat cannot be fully dissipated and leaf tissue damage may occur. Some antitranspirants may clog the stomates, reducing photosynthesis as well as water loss. Their use may therefore be most beneficial over short periods of time.

DRAINAGE

A poorly drained planting site can cause decline of most tree species. Excess water in the root zone suffocates and kills roots (Figure 4.8). It is important to consider the grade and drainage flow pattern during construction or landscape development. Attempts should be made to avoid creation of low spots where water could pool or drain slowly. If the site is not already developed, adjusting the grade is the first choice to avoid drainage problems. If a developed site is known to have poor drainage, drain tiles can be installed prior to planting to carry excess water away from the planting site. Drain tiles are made of clay, concrete, or plastic, although plastic is most commonly used for landscape purposes. The depth and spacing of the tiles depends on soil and planting conditions. Drain lines should be placed above any deep, impervious layer or hardpan, and they should slope away from the plantings at the rate of ¼ inch per linear foot. Tiles are normally placed on sand or fine gravel about 3 feet below the soil surface. The holes in the drainage tiles should be placed facing down so that water enters the drain from below, after most of the large particles have settled out. Wrapping drain tiles in geotextile fabric can help prevent soil penetration.

Drainage problems in urban landscape sites are often not recognized until after planting is complete. If drain tiles cannot be installed, surface drainage may be improved by changing the grade or trenching. However, care must be taken to avoid damaging the root systems of existing trees.

In soils in which water infiltration is slow, steps can be taken to avoid standing water. Slow water application and root-zone aeration can help prevent puddling and runoff. Irrigation should not be applied faster than water can infiltrate. Water penetration and aeration can be improved by drilling holes 6 to 18 inches deep and backfilling with an appropriate material such as sand or pea gravel.

There are limitations to the effectiveness of maintenance practices in controlling drainage. Careful plant selection is important in choosing trees that will adapt to wet soil conditions.

Figure 4.8 Drainage is an important consideration in water management. Trees planted in poorly drained soils are subject to root disease and poor root development. It is best to check drainage before trees are planted and make modifications, if necessary.

Chapter 4 Workbook

1. True/False—Infrequent, deep soakings of trees and shrubs are preferable to frequent, shallow waterings.
2. The most beneficial times to irrigate plants are _______ ________ or ________ ____________.
3. The ________________ rate is the rate at which water soaks into the soil.
4. Clay soils generally have a greater _______ ____________ _______________ than sandy soils, but the water percolates through clay soils more slowly.
5. Disadvantages to sprinkler irrigation are high application rates and surface _______________ created by the water hitting the soil surface.
6. True/False—In most instances, it is preferable to minimize water on the foliage of plants.
7. Name two advantages and two disadvantages of drip irrigation.

Advantages	**Disadvantages**
a.	a.
b.	b.

8. ______________ ________________ is designed to maintain plants during periods of reduced rainfall, supplying only enough water to maintain a desired plant quality.
9. A measure of the rate of water use by plants and evaporation from soil is known as __________________.
10. Name five benefits of using mulches around trees.
    a.
    b.
    c.
    d.
    e.
11. _________ _____ can occur in trees when mulch or soil is placed against the bark on the trunk of the tree.
12. _________________ are materials sprayed on plants to reduce water loss through transpiration.
13. A ______________________ is a soil moisture sensor used to monitor soil wetness or dryness.
14. True/False—Drought problems can be especially severe following periods of moderate to high soil moisture conditions.
15. Name three tree health problems associated with flooding.
    a.
    b.
    c.

CHALLENGE QUESTIONS

1. What are some of the methods commonly used to estimate how often and how much to irrigate?

2. How does soil salinity occur? What are the irrigation practices that should be used to minimize soil salinity?

3. In what ways can antitranspirants reduce water loss through transpiration? What are the limitations to their use?

4. Compare and contrast the effects and symptoms of drought versus flood conditions.

SAMPLE TEST QUESTIONS

1. When irrigating trees,
   a. infrequent, deep soakings are preferable to frequent, shallow waterings
   b. the most beneficial and efficient time to water is mid-afternoon at peak sunlight
   c. the foliage should be kept wet at night to reduce transpiration
   d. keeping the soil moist at the root flare reduces girdling root formation

2. Sandy soils
   a. have a greater water-holding capacity than clay soils
   b. have a higher infiltration rate than clay soils
   c. do not ever reach field capacity because drainage is good
   d. all of the above

3. A soil is at field capacity when
   a. it is completely saturated
   b. the permanent wilting point has been reached
   c. gravitational water has drained away
   d. there is no water available to the roots

4. With sprinkler irrigation,
   a. less frequent waterings of longer duration will reduce the likelihood of salt buildup
   b. water hitting the soil surface can lead to crusting
   c. application rates that exceed absorption can lead to runoff
   d. all of the above

5. In sites where poor drainage can be a problem, which of the following **will not** help ensure plant survival?
   a. careful plant selection and planting shallow
   b. improvement of surface drainage by changing the grade or trenching
   c. the installation of drain tiles
   d. placing gravel in the bottom of the planting hole and backfilling with a sandy soil

Other Sources of Information

(See pages v–vi for complete bibliographic information.)

Harris et al., 1999. *Arboriculture: Integrated Management of Landscape Trees, Shrubs, and Vines.*
Kramer, 1969. *Plant and Soil Water Relationships: A Modern Synthesis.*

CHAPTER 5

TREE NUTRITION AND FERTILIZATION

CHAPTER 5 TREE NUTRITION AND FERTILIZATION

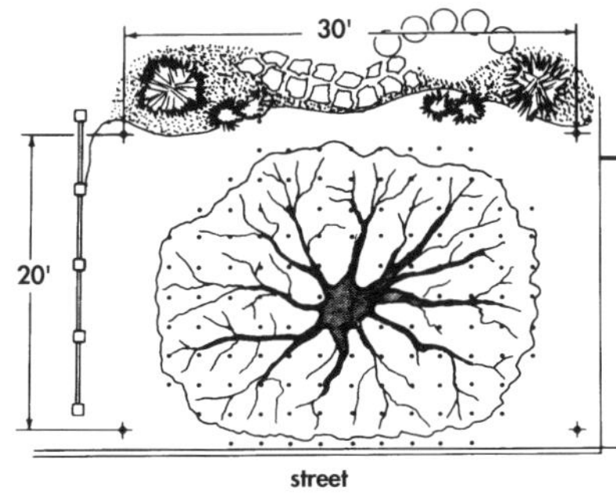

CHAPTER 5
TREE NUTRITION AND FERTILIZATION

Objectives

1. Understand the reasons for fertilizing urban trees.
2. Become familiar with the essential elements that a tree requires and how these elements are absorbed.
3. Explain the various types of fertilizer and the advantages and disadvantages of each.
4. Describe the different methods of fertilizer application and the advantages and disadvantages of each.
5. Discuss the problems that can be associated with excessive fertilization and with leaching.
6. Explain why determining nutritional requirements and availability is the first step prior to fertilization recommendations.

Key Terms

ANSI A300 standards	essential elements	inorganic	organic	soil analysis
complete fertilizer	fertilizer analysis	leaching	prescription fertilization	subsurface application
controlled release	fertilizer burn	liquid injection	secondary nutrients	surface application
drill-hole method	foliar analysis	macronutrients	slow-release fertilizer	water-insoluble nitrogen (WIN)
drip line	foliar application	microinjection		
	implants	micronutrients		

INTRODUCTION

Trees require certain **essential elements** to function and grow (Figure 5.1). An essential element is a chemical element that is involved in the metabolism of the tree or necessary for the tree to complete its life cycle. For trees growing in a forest site, these elements are usually present in sufficient quantities in the soil. Landscape trees or urban trees, however, may be growing in soils that do not contain sufficient available elements for satisfactory growth and development. Leaves and other plant parts are regularly raked up and removed, disrupting nutrient cycling and the deposition of organic matter back into the soil. In addition, the population and activity of the microorganisms that help to break down materials and release elements are often decreased. It may be necessary to fertilize or to adjust the soil pH to increase available elements.

Fertilizing a tree with nitrogen can increase growth, reduce susceptibility to certain diseases and pests, and can, under certain circumstances, help reverse declining health. However, if the fertilizer is not needed or not applied wisely, it may not benefit the tree at all—and may adversely affect the tree. Trees making satisfactory growth and not showing symptoms of nutrient deficiency may not require fertilization. Trees growing in turf that is routinely fertilized, or where clippings are returned to the soil, may not require supplemental fertilization. Nitrogen can unnecessarily increase growth, requiring more pruning. It is important to recognize when a tree needs supplemental fertilizer, which elements are needed, and when and how the fertilizer should be applied.

Tree fertilization should be done in accordance with **ANSI A300 standards** developed by the Accredited Standards Committee on Tree, Shrub, and

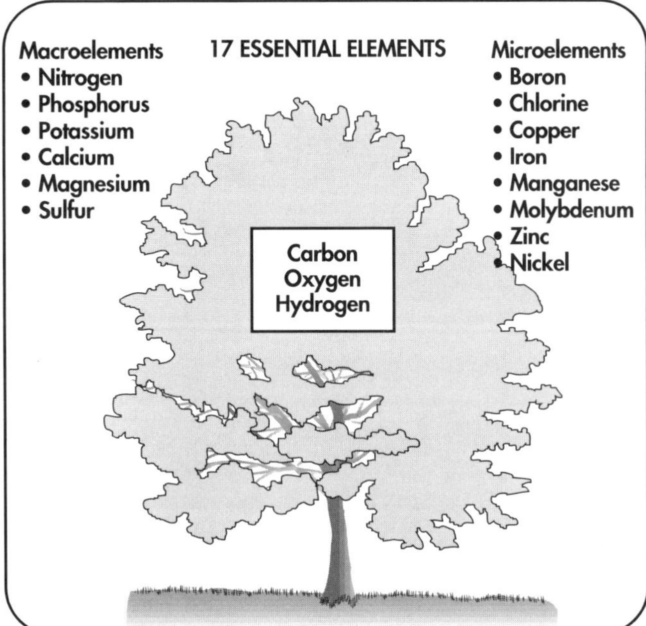

Figure 5.1 Trees require 17 essential elements.

Other Woody Plant Maintenance Operations. These standards are intended for use in preparing specifications for tree maintenance. Specifications for tree fertilization should clearly define the objectives and should specify the type of fertilizer, its rate and timing, and the method and location of application.

TREE REQUIREMENTS

Trees take up essential elements dissolved in water through their roots. Each of the essential elements has a specific role in a plant and cannot be replaced by another. Trees require certain elements, known as **macronutrients**, in relatively large quantities. The most important of these macronutrients is nitrogen (N), because it is most often limiting. Nitrogen is a constituent of proteins and chlorophyll, and it is critical to photosynthesis and other plant processes.

Under natural conditions, soil nitrogen comes largely from organic matter, as well as from the atmosphere. Soil organisms decompose the organic matter, releasing nitrate ions (NO_3^-) and ammonium ions (NH_4^+), which, in turn, can be further converted to nitrate ions. The positively charged ammonium ions are adsorbed to soil particles, while the negatively charged nitrate ions are free in the soil water to be picked up by plant roots or leached down through the soil. Much of the nitrogen in the soil can be lost due to leaching or volatilization (the return of nitrogen to the atmosphere in its gaseous state). Removing leaf litter and other natural sources of nitrogen can disrupt the cycling of nitrogen in the soil. Nitrogen deficiencies are most common in sandy soils low in organic matter, especially where irrigation is heavy.

Nitrogen deficiency shows up as reduced growth, smaller leaves, and yellowing (chlorosis) of the leaves, especially the older leaves. Sometimes the newer, developing leaves appear greener because nitrogen is somewhat mobile within plants, allowing it to be directed toward new growth. However, these symptoms could also be due to a variety of other problems that affect root health and element uptake. Because nitrogen is the element most likely to be deficient in trees, fertilization specifications usually focus on supplementing it.

In addition to nitrogen, phosphorus (P), potassium (K), and sulfur (S) are also required in relatively large quantities, but these elements are usually present in adequate amounts in the soil for trees and large shrubs. **Secondary nutrients**, which are required in moderate quantities, include magnesium (Mg) and calcium (Ca).

Other elements, known as **micronutrients**, are required in lesser quantities. Although these elements are not required in large amounts, a deficiency of any one can have profound effects on the health of the tree. For example, iron chlorosis is a condition that results when a tree is not absorbing sufficient quantities of iron or is unable to use the iron that has been absorbed. Young leaves are small and chlorotic (yellow), often with green veins, while older leaves tend to be darker green. Iron deficiency can eventually kill a tree. Like iron, manganese and zinc may at times be deficient in a tree. Chlorosis symptoms due to manganese deficiency are similar to those of iron deficiency. Trees deficient in zinc may have small leaves and stems that fail to elongate (rosetting). The remaining elements—molybdenum, copper, chlorine, boron, and nickel—are less likely to be deficient (Figure 5.2).

FERTILIZER

Fertilizers are available in many forms and combinations. A **complete fertilizer** is one that contains nitrogen, phosphorus, and potassium. The **fertilizer analysis**, listed on the container, gives the composition of the fertilizer expressed as a percentage by weight of total nitrogen (N), available phosphoric acid (P_2O_5), and soluble potash (K_2O), always listed in the same order (Figure 5.3). For example, a fertilizer with an analysis of 10-6-4 contains 10 percent nitrogen, 6 percent phosphorus, and 4 percent potassium. A 100-pound bag of this fertilizer would contain 10 pounds of nitrogen. For trees, a complete fertilizer is usually not necessary because nitrogen may be the only limiting element.

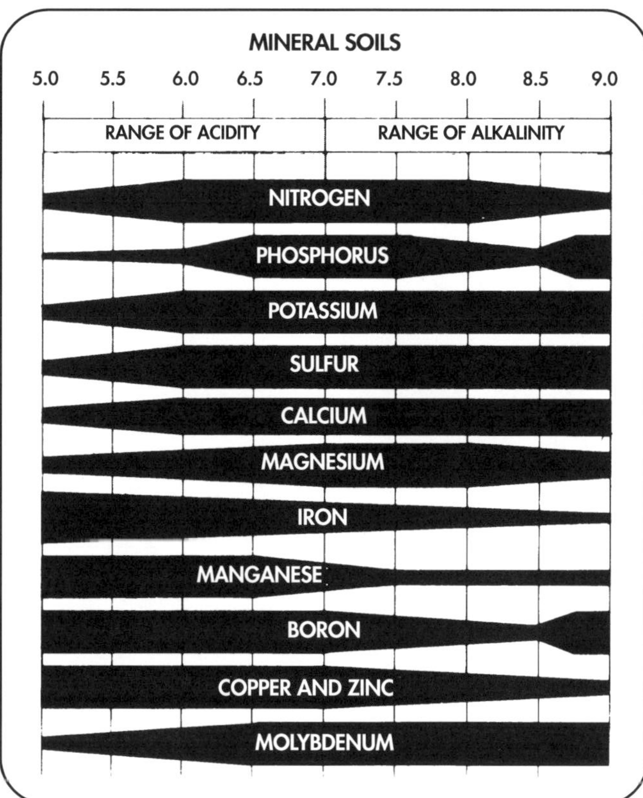

Figure 5.2 Relative availability of essential elements to plant growth at different pH levels for mineral soils.

Figure 5.3 The fertilizer analysis of this fertilizer is 20-6-4. The ratio is 10:3:2. Because this is a 40-pound bag with 20 percent nitrogen, there are 8 pounds of actual nitrogen in this bag.

Fertilizers are available in either an **organic** or **inorganic** form. Inorganic fertilizers release their elements relatively quickly when dissolved in water. The available nutrients are in the form of inorganic ions that are absorbed by oppositely charged sites on the root membrane. These same ions are also responsible for plant "burn" (when in excess concentration) by raising the osmotic pressure of soil solution and drawing water out of the roots.

Organic fertilizers also release inorganic ions, but more slowly as the molecules are hydrolyzed or decomposed in the soil. Organic fertilizers are composed of carbon-based molecules and can be either synthetic or natural. Examples of synthetic organics are urea formaldehyde and isobutylidene diurea (IBDU). Examples of natural organics are manures, sewage sludge, blood, and bone meal. Urea, which contains a single carbon atom, is technically an organic. However, it is not recognized as such in the industry because of how rapidly it solubilizes and releases nitrogen ions in water.

Roots can absorb most elements in the form of inorganic ions only, whether coming from an organic or inorganic source. An advantage of organic fertilizers is that they must be converted to inorganic ions before absorption; therefore, they are not leached as readily from the soil. One advantage of inorganic fertilizers is that solubility is less affected by temperature so that the rate of availability is more uniform.

Often it is desirable to use a **slow-release fertilizer.** Slow-release fertilizers release the fertilizer, usually nitrogen, over an extended period of time. This reduces the amount of fertilizer that may be leached and also reduces salt and fertilizer "burn" problems. If slow-release fertilizer is used, more nitrogen can be applied at one time.

One type of slow-release fertilizer has an insoluble coating that extends the release period. An example is sulfur-coated urea. Other slow-release fertilizers are synthetic or natural organics whose release is determined by the carbon structure. Examples are urea-formaldehyde (synthetic) or manures (natural). For long-term release of nitrogen, trees should be fertilized with either a slowly soluble source of at least 50 percent **water-insoluble nitrogen (WIN)** or a slow-release source with a 20 to 30 percent dissolution rate.

Rates

The rate that the fertilizer should be applied depends on the age, health, and species of the tree, the form of the fertilizer, the method of application, and the site conditions. Application rate is a controversial subject, and some researchers are beginning to recommend lower rates than have commonly been used. The general recommendation for slow-release fertilizers is 2 to 4 pounds of actual nitrogen per 1,000 square feet of root area. Quick-release fertilizers should be used only when the objectives of fertilization cannot be met with slow-release fertilizer. Rates should be between 1 and 3 pounds of actual nitrogen per 1,000 square feet per application and should not exceed 4 pounds of actual nitrogen annually. If the fertilizer is being broadcast over turfgrass, no more than 3 pounds of soluble nitrogen per 1,000 square feet should be applied to avoid damaging the turfgrass. It is necessary to thoroughly water the area to dissolve the fertilizer after application. Water is critical to uptake of essential elements.

Timing

Regardless of when fertilizer is applied, it may not be readily absorbed or utilized until growth begins in the late winter/early spring. Uptake and metabolic demand are low in the dormant season, and some of the more soluble forms of nitrogen may leach from the soil before they can be utilized. Studies show that nitrogen uptake peaks during the spring and summer when metabolic need is greatest.

The most limiting factor in fertilizer uptake is water availability. Fertilization studies on mature trees show that response is greatest when moisture levels are high. The frequency of application depends on the tree and on soil conditions. Recall that clay soils have a higher cation exchange capacity, which means they can readily attract, adsorb, and exchange positively charged minerals. In sandy soils, smaller amounts of fertilizer may need to be applied more frequently than in clay soils because of a greater possibility of leaching. Another option for sandy soils is to use a slow-release fertilizer with a high level of WIN. Fertilizer uptake is greatest during periods of active root growth, so applications of fast-release fertilizers are most effective spring through fall. Slow-release fertilizers can be applied at any time the ground is not frozen and when soil moisture is adequate.

Application Techniques

Applying fertilizer to the soil surface is the easiest and least expensive method of fertilizing trees. In **surface application,** the fertilizer is broadcast over the soil surface using a spreader calibrated to apply the desired amount of nitrogen per 1,000 square feet (100 square meters) (Figure 5.4). Following application, the area should be thoroughly watered to dissolve the fertilizer and wash it off the grass and into the soil. Care must be taken not to apply soluble nitrogen in rates great enough to burn the grass.

There are several advantages to surface application of fertilizer to trees, and numerous studies have shown it to be an effective means of providing nitrogen to trees. It requires the least amount of time for application and does not require sophisticated equipment. The nitrogen is available in the upper 12 inches of soil, where most of the tree's actively absorbing roots are located. Because soluble nitrogen tends to leach with the downward-draining water, more may be available where the uptake is greatest. Some methods of "deep root" fertilization may place the fertilizer below the zone where the fibrous roots are located. Surface application should not be employed, however, where runoff is a problem.

Figure 5.4 Handheld spreaders are one way to surface-apply fertilizer.

When the tree is growing in a lawn, turfgrass will compete significantly for available nitrogen. Although tree roots will get some benefit, turf roots will outcompete tree roots in absorbing essential minerals because turf roots respond more quickly to fertilization. In these cases, some arborists prefer to use one of the **subsurface application** techniques. Arborists should keep in mind, however, that the majority of the fine, absorbing tree roots are located in the upper inches of the soil. As always, it is important to determine whether supplemental fertilization is required.

Subsurface application is designed to place fertilizer below the surface and below the majority of the turfgrass roots. The **drill-hole method** uses granular fertilizer (Figure 5.5). Holes 2 to 4 inches in diameter, spaced 12 to 36 inches apart, are drilled in the soil around the tree in concentric circles or in a grid pattern (Figure 5.6). It is best to begin the holes several feet out from the trunk to avoid damaging the buttress roots. The holes should extend at least to the **drip line**, although property lines, pavement, and buildings usually restrict the application area. Fertilizer holes should not be poked or punched into the soil because it causes compaction and glazing around the holes and may restrict distribution and uptake of the fertilizer.

Figure 5.5 Drill-hole fertilization.

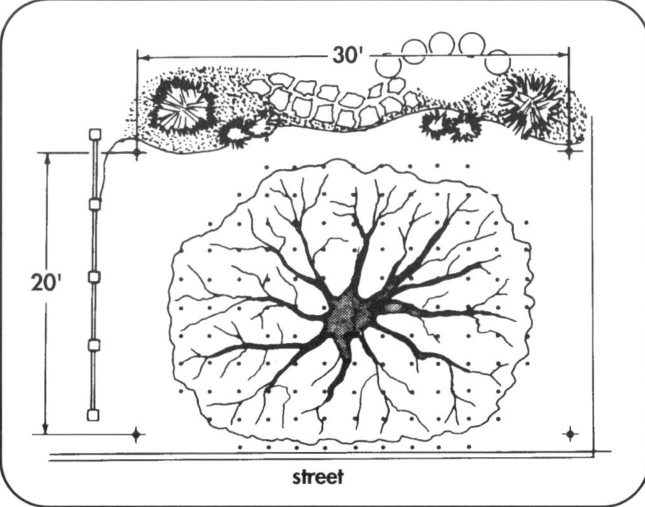

Figure 5.6 Because the root system does not stop at the drip line, it may be beneficial to fertilize beyond. In residential areas, property lines or paved surfaces may limit the application area.

The holes should be 4 to 12 inches deep, depending on soil type and other site conditions. After the amount of fertilizer required is calculated according to specifications, the fertilizer is uniformly distributed evenly in the holes. Remember that the nitrogen washes down, and if placed too deeply nitrogen will not be available to the roots. The fertilizer should not fill the holes within 2 inches of the surface. Sometimes the holes are backfilled with a soil amendment such as peat moss, perlite, crushed stone, pea gravel, or sand. One advantage of drill-hole fertilization is the partial aeration of the soil.

A third application technique is the **liquid injection** method. Fertilizer dissolved or suspended in water is injected under hydraulic pressure into the soil using a soil injector (Figure 5.7). The hole spacing and distribution are the same as with the drill-hole method: holes 4 to 12 inches deep and 12 to 36 inches on center. The advantages of this method are better distribution of fertilizer and the injection of water into the root zone. Often the factor limiting fertilizer uptake is lack of available water. Liquid injection of fertilizer reduces this problem.

Figure 5.7 Liquid injection fertilization.

Both the drill-hole method and the liquid injection method of fertilization can lead to dark, vigorous patches of lawn under the tree around each distribution site. Some arborists use subsurface methods in conjunction with broadcast fertilization. This eliminates uneven discoloration in grass and provides the advantages of each application technique. Total fertilizer applied should not exceed standard recommendations, however.

Foliar application, another method of fertilization, is sometimes employed to correct minor element deficiencies. For example, chelated iron sprays may be used to provide a rapid, although temporary, treatment of iron chlorosis. This method should not be considered a sufficient means of providing all the necessary mineral elements in the amounts necessary for adequate tree growth. It is, in fact, a rather ineffective means of applying nitrogen.

Micronutrient spray applications are most effective when made just before a period of active growth. Response time varies, but one or two applications per year will usually be effective. However, not all plant species respond to foliar treatments.

Implants and injections are two techniques employed to introduce chemicals directly into the xylem of trees. Implants, such as capsules and some **microinjections,** rely on the transpirational stream to move materials systemically within the xylem (Figure 5.8). Some injection techniques involve supplemental pressure to reduce injection time. Implants and injections have been used successfully to introduce essential elements into trees to treat micronutrient deficiencies, but they are not practical for supplementing macronutrients because of the volume required. As with foliar sprays, these techniques are most successfully used to correct minor element deficiencies. Tree uptake and distribution is best achieved when the tree is actively transpiring due to the negative pressure in the xylem. This treatment should never be used on drought-stressed trees.

Technique is important when using tree injections. To minimize wounding, small, clean holes should be drilled close to the base of the tree at the root flare. Insertions should be made soon after drilling to reduce the chances of a blockage inhibiting uptake. Implants or injections should be placed below the bark and cambium, directly into the first rings of the xylem. Injection wounds directly above or below other wounds may cause severe injury. Injection wounds from previous applications should not be reused. Trees less than 4 inches in diameter should not be treated with implants or injections.

Figure 5.8 Microinjection application.

Many experts are hesitant to recommend implants or injections because chemical toxicity may be severe, and damage to the cambium and xylem may result. Although treatments using systemic injections may yield remarkable results, repeated use can cause wounds to coalesce. Because of this, implant or injection treatments should never be made less than one year apart, and a minimum of two years is preferred. These techniques can be a temporary treatment, but they are not recommended for routine use. They should be applied only when soil application techniques are impractical or ineffective.

SOIL ADDITIVES

A number of products are available and marketed as soil additives. Many claim to stimulate growth or improve plant health. Soil additives range from various forms of organic matter to vitamins, mineral elements, and seaweed. Unfortunately, little research is available on the efficacy of these products, and there is little evidence of their value.

One popular group of additives are soil inoculants. Recognizing the important role of mycorrhizae in water and mineral absorption, researchers have sought to encourage mycorrhizal development in tree roots. It should be clarified, however, that mycorrhizae are a symbiotic relationship between tree roots and a fungus. It is not possible to introduce mycorrhizae into the soil. Rather, the goal is to introduce the fungal inoculants, which will in turn form mycorrhizae. Mycorrhizae are influenced by soil pH, light, minerals in the soil and plant, and other competing mycorrhizae and soil microbes. Inoculation has been successful in very poor soils where fertility and microbe populations are low or nonexistent. In more favorable soils, little benefit has been shown, and the mycorrhizal spores may serve as a food source for other microbes. Research results on the value of these products have been mixed, and certainly more research is needed.

SALT PROBLEMS

Recommended fertilizer rates are based on achieving good tree response while avoiding **fertilizer burn**. Fertilizers are actually salts of different chemicals that are essential for plant growth and health. Fertilizer "burn" is the same type of injury that can be caused by common salt (sodium chloride) used for de-icing. Like common salt, excessive soluble fertilizer in the root zone can actually draw moisture out of the roots (Figure 5.9). Symptoms range from wilting and marginal burning of the foliage to death of the plant. Many fertilizer labels indicate the salt index, which is a relative scale that indicates the likelihood to cause burning. A fertilizer with a salt index below 50 is usually preferred. The use of slow-release fertilizers can minimize salt injury.

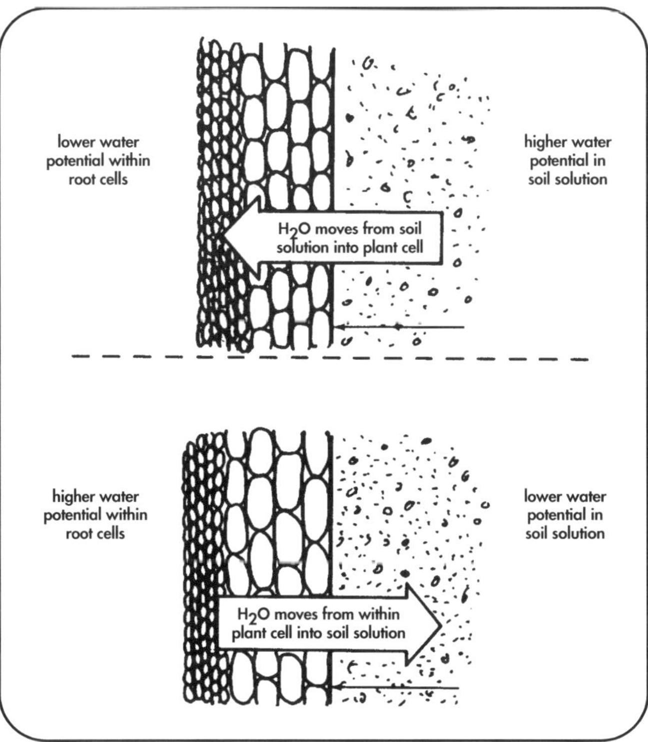

Figure 5.9 Excessive soluble salts in the root zone can reverse the osmotic process, drawing water out of the roots.

LEACHING AND UPTAKE PROBLEMS

Leaching is the washing of chemicals down through the soil. Nitrogen and other water-soluble elements tend to leach from the root zone. Not only does this make them unavailable to the tree, but they can also pollute groundwater, lakes, and streams. To minimize leaching, apply only the amount of fertilizer needed, use organic or slow-release forms, and avoid over-irrigating in sandy soils.

Another problem that may reduce availability of certain elements is pH. Soil pH can influence solubility of certain elements by causing them to form insoluble salts. An example is the micronutrient iron. If the soil is too alkaline, iron solubility is reduced and iron is unavailable to the tree. Thus, adding iron to the fertilizer will provide only temporary relief because the added iron will soon become unavailable. For this reason, it is advisable to perform foliar and soil analyses before treatment.

SOIL AND FOLIAR ANALYSES

The most accurate way to determine a tree's element needs is to obtain laboratory analyses of the soil and leaves. A **soil analysis** can give information about the presence of essential elements, soil pH, organic matter, and cation exchange capacity. The pH and salt content (especially in arid regions) are most important for recommending treatments. When taking a soil sample, small cores should be taken from representative locations of the entire area or root zone. These cores should be mixed together. This procedure gives results that are averaged over the entire area, avoiding pockets where levels may vary. Any soil test is only as good as the sample, so it is important to collect a representative sample of the site.

A soil analysis has greater value if done in conjunction with a **foliar analysis**. Because the roots of trees grow over a vast area and at various depths, soil analysis alone may not yield accurate results for certain soil elements. Also, there is still debate about what levels of various elements are critical for tree growth—and required levels vary by tree species. Thus, interpretation may be difficult. One way to facilitate interpretation is to compare samples from healthy and symptomatic trees of the same species. Leaf samples taken from all over the tree, dried, and analyzed, may help diagnose certain deficiencies or toxicities. However, a soil analysis or foliar analysis alone can be misleading. It is possible for certain minerals to be deficient in leaves but plentiful in soil and unavailable due to the soil pH.

Recommendations for fertilization must be made based on element requirements and availability in the soil. Although some fertilization can increase growth, it is not always beneficial. Excess fertilization can injure trees and reduce tolerance to certain stress problems. Studies now show that nitrogen fertilization can trigger a tree's energy allocation toward growth, sometimes at the expense of defense. Resistance to certain pests can decrease in rapidly growing trees, and pest problems may increase on regularly fertilized trees. And, although fertilization is often recommended to help give ailing trees a boost, trees stressed by drought, poor soil aeration, and some diseases may not respond to fertilization.

Fertilization is just one plant health care treatment available to arborists. Professionals must understand its proper and judicial use. Current practices are moving toward what is known as **prescription fertilization**, based on element analysis. This is an apt term, keeping in mind the medical saying, "prescription without diagnosis is malpractice."

Chapter 5 Workbook

1. Trees take up essential elements, dissolved in ________, through the roots.
2. _________________ are elements required by trees in relatively large quantities.
3. The macronutrient ________________ is a constituent of chlorophyll and, if deficient, can cause reduced growth and yellowing of the foliage.
4. Yellowing between the leaf veins is called ____________ and may be the result of mineral deficiencies.
5. A complete fertilizer contains _____________, ____________, and ________________.
6. The ____________ ____________, listed on the container, gives the relative percentage of nitrogen, phosphorus, and potassium.
7. A 50-pound bag of 20-10-5 fertilizer contains ______ pounds of actual nitrogen.
8. Manure, bone meal, and sewage sludge are _____________ forms of fertilizer.
9. If fertilizer "burn" or leaching are potential problems, it may be desirable to use a __________ _____________ fertilizer.
10. The general recommendation for fertilizer rates for trees using a slow-release fertilizer is ______ pounds (actual nitrogen) per 1,000 square feet of root area.
11. The most important factor for good uptake of fertilizer elements is adequate ________________.
12. True/False—Surface application of fertilizer is relatively inexpensive and makes the fertilizer available in the upper few inches of soil.
13. The drill-hole method of fertilizer application has the added advantage of _______________ the soil.
14. What is the biggest problem with deep root fertilization? ____
15. Foliar application of fertilizer is sometimes used to correct ___________________ deficiencies.
16. Name two limitations to trunk implants and microinjections.

    a.

    b.
17. Wilting, marginal burning, and dieback may be symptoms of excess ____________ ______ in the root zone.
18. _______________ is the washing out of chemicals down through the soil.
19. Fertilization recommendations should be based on ______________ ___________________.
20. Two of the most important levels measured in a soil analysis are the _____________ and the salt levels (especially in arid regions).

MATCHING

___ micronutrients	A. recommended rate for trees
___ 10-6-4	B. slowly soluble, synthetic organic
___ ureaformaldehyde	C. interveinal foliar yellowing
___ complete fertilizer	D. copper, chlorine, boron
___ 2 to 4 pounds/1,000 square feet	E. nitrogen, phosphorus, and potassium
___ chlorosis	F. fertilizer analysis

CHALLENGE QUESTIONS

1. Explain how a nutrient can be plentiful according to a soil analysis, yet be deficient in a tree.

2. If a fertilizer with an analysis of 10-0-0 costs 15 dollars for a 60-pound bag, what will be the fertilizer cost for fertilizing 2,000 square feet of root area at the rate of 2 pounds per 1,000 square feet?

3. What are the advantages to using slow release forms of nitrogen?

4. Compare and contrast the various methods of fertilizer application, including surface broadcast, drill-hole method, liquid injection, foliar spray, and implants or injections.

SAMPLE TEST QUESTIONS

1. An 80-pound bag of 10-6-4 fertilizer contains how many pounds of actual nitrogen?
   a. 6
   b. 8
   c. 10
   d. 4
2. A complete fertilizer contains
   a. all 17 essential elements
   b. nitrogen, phosphorus, and potassium
   c. organic and inorganic nitrogen
   d. equal amounts of N, P, and K
3. A tree may not respond immediately to fertilizer application if
   a. a slow-release fertilizer was applied
   b. there is inadequate soil moisture
   c. the tree is suffering from collar rot
   d. all of the above

4. A soil test may not identify a nutrient deficiency problem in a plant because
   a. the tests are not reliable
   b. the nutrient content can change after collecting
   c. the soil may contain adequate nutrients but something may be inhibiting uptake
   d. no one knows which levels of nutrients in soils are adequate

5. Sulfur-coated urea or ureaformaldehyde is sometimes included in nitrogen fertilizers because
   a. slow-release forms of nitrogen are sometimes desired
   b. the soil pH must be adjusted to optimize nutrient uptake
   c. the soil's buffering capacity may inhibit nitrogen absorption
   d. urea is a good natural source of phosphorus

Other Sources of Information

(See pages v–vi for complete bibliographic information.)

ANSI A300. *Standard Practices for Tree, Shrub, and Other Woody Plant Maintenance, Part 2: Fertilization.*
Harris et al., 1999. *Arboriculture: Integrated Management of Landscape Trees, Shrubs, and Vines.*

CHAPTER 6
TREE SELECTION

CHAPTER 1 TREE BIOLOGY

CHAPTER 2 TREE IDENTIFICATON

CHAPTER 3 TREE/SOIL RELATIONS

CHAPTER 4 WATER MANAGEMENT

CHAPTER 5 TREE NUTRITION AND FERTILIZATION

CHAPTER 6 TREE SELECTION

CHAPTER 7 INSTALLATION AND ESTABLISHMENT

CHAPTER 8 PRUNING

CHAPTER 9 TREE SUPPORT AND PROTECTION SYSTEMS

CHAPTER 10 DIAGNOSIS AND PLANT DISORDERS

CHAPTER 11 PLANT HEALTH CARE

CHAPTER 12 TREE ASSESSMENT AND RISK MANAGEMENT

CHAPTER 13 TREES AND CONSTRUCTION

CHAPTER 14 SAFETY

CHAPTER 15 CLIMBING AND WORKING IN TREES

CHAPTER 6
TREE SELECTION

Objectives

1. Discuss the benefits of trees to the landscape and the environment.
2. Describe the site characteristics that should be considered prior to tree selection.
3. Explain the characteristics of a tree that must be considered when selecting a species.
4. Explain what to look for in selecting healthy, vigorous planting stock.

Key Terms

acclimation
adaptability
design criteria
growth rate
hardiness
introduced species
landscape function
native species
naturalized species
pest resistance
site analysis
site considerations

INTRODUCTION

Trees are one of our most important landscape features. They beautify and enrich our lives throughout the year with their variety of forms, flowers, fruits, seasonal leaf colors, and bark interest. They provide oxygen and cooling shade, and they filter out some pollutants from the air. Trees deflect noise and wind, block unwanted views, and protect the soil from erosion. They can also reduce rain runoff by capturing and holding rainwater before it reaches the ground. Too often, however, we take trees for granted, not stopping to think that the benefits trees provide can be the result of careful planning and selection. In fact, if a tree is not properly matched to the landscape site, the tree can become more of a liability than an asset. Choosing the right tree for a particular site is one of the most important decisions to ensure long-term benefits, beauty, and satisfaction.

MATCHING TREE AND SITE

The first important part of tree selection is matching the tree with the site. Every tree species has certain cultural requirements including light, water, soil conditions, growing space, and others. Each planting site has environmental characteristics such as temperature extremes and light levels that can limit which plants will thrive (Figure 6.1). In addition, most trees are planted to fulfill a particular function in the landscape. For example, trees may be planted to create shade for a backyard patio. So, it is important to choose a tree species or cultivar that will fulfill the desired function and be capable of growing in the cultural and environmental conditions at the site (Figure 6.2).

Figure 6.1 Keep in mind the mature size of the tree when selecting a species for a site.

Figure 6.2 Consider the desired function for each plant in the landscape setting.

With such a variety of plant and **site considerations**, it is necessary to set priorities in selecting a tree. The highest priorities are those that will affect the survival of the tree, such as ability to irrigate, drainage, pH adaptability, light, and moisture needs. Sometimes the functional goals cannot be met entirely because of site limitations. For example, if a screen is desired, an evergreen species might be selected because it would provide year-round screening. However, if shade is a limiting factor, a deciduous species might be selected because there are few evergreens that grow well in the shade.

Adaptability is a tree's genetic ability to adjust to different conditions. Some species are more adaptable than others. For example, red maples *(Acer rubrum)* are native from Florida to Canada, while palms will grow in warm climates only. **Acclimation** is the process by which a given tree adapts to its environment. This usually involves physiological or morphological changes. For example, some trees acclimate to shade by developing larger, thinner leaves. Trees with variegated or colored foliage tend to lose variegation or color when grown in the shade, maximizing their photosynthetic capacity.

While it is possible for some trees to become acclimated to certain site conditions such as partial shade and low moisture, a tree will have a much greater chance of survival if planted in an appropriate site. Some books and software provide lists of trees adaptable to certain site conditions such as shade, wet areas, and urban situations. Although such lists may be very helpful, it is important to keep in mind that many factors influence survival. For example, baldcypress *(Taxodium distichum)* is considered to be tolerant of wet soils. However, if the tree has been grown in an upland nursery area, it may not do well if transplanted into a wet site. Looking again at our red maple example: They have a wide native range, but trees selected from a southern seed source may not be hardy in a northern climate, and vice versa.

Computer-aided plant selection programs have made the plant selection process a little easier. These programs allow the user to generate a list of trees that match site and plant characteristics chosen by the user. For example, the software could generate a list of trees that are native to a region, are drought tolerant, are less than 35 feet tall, and have showy, white flowers.

Site Considerations

Many competent professionals do a **site analysis** before a landscape plan is designed. The site analysis records pertinent existing site conditions that will affect plant selection. In addition, the functional goals of the design are outlined. This helps the designer select plants that are appropriate for the site.

There are many site characteristics that must be taken into consideration in plant selection (Tables 6.1 and 6.2). Plants may be limited by growing space, water availability and drainage, soil texture and pH, light levels, weather, and site use or function. Trees can suffer if one or more of these factors are incorrect or incompatible.

The amount of growing space available is important when selecting the type of tree to plant in a particular site. The area above- and belowground must be large enough to allow the tree to reach its mature height, branch spread, and trunk diameter without interfering with surrounding objects. It is easy to forget that many trees can grow more than 100 feet in height if conditions are favorable. Many trees grow wider than tall.

Large trees can be wonderful assets, especially for shade, as long as they have enough room to mature. However, tall-growing trees should never be planted beneath utility wires (Figure 6.3). Trees without sufficient belowground growing space can interfere with sidewalks and curbs and will likely be stressed (Figure 6.4). Similarly, by knowing the mature branch spread, one can avoid planting a tree too close to a building where it will cause damage to the roof or siding, or too near an intersection where it may block visibility.

The climatic or environmental conditions of the region must also be considered. This includes not only the climate of the geographical zone but also the microclimate of the planting site, which can be affected by buildings, topography, pavement, and other surroundings.

Table 6.1 Site characteristics.

Climate
- Hardiness zone
- Rainfall/snowfall
- Sunlight/other lighting
- Prevailing winds
- Exposure

Soil
- pH
- Soil texture
- Bulk density
- CEC
- Nutrient analysis
- Soil volume

Planting site
- Buildings and other structures
- Paved surfaces
- Plans for future development
- Overhead and underground utilities
- Intended use of site
- Intended function of plant

Other plantings
- Trees and shrubs
- Planting beds
- Groundcovers/flowers
- Turf

Maintenance to be provided
- Irrigation
- Post-planting care
- Ongoing maintenance

Other requirements

Light levels may significantly affect plant growth and survival. We often think of excessive shade as a problem to some trees because many trees cannot survive in a heavily shaded site. Their branches grow long and spindly, and they may drop some leaves. Bright light or reflected light and heat from buildings and paved areas can also cause severe problems if a tree cannot adapt to those conditions (Figure 6.5). Some trees will develop scorched foliage and may wilt in the bright afternoon sun. Others are adapted to bright light and heat and have small, thick foliage, and sunken stomata to reduce transpiration.

The soil conditions of the site must also be considered. It may be advisable to get a soil analysis to determine the texture, pH, soluble salts, and nutrient levels of the soil. One of the most important aspects of the soil is its water-holding capacity and

Table 6.2 Tree characteristics.

Hardiness

Growth habit/form

Size
- Height
- Spread
- Root zone requirements

Attributes
- Showy flowers
- Attractive fruit
- Attractive bark
- Interesting foliage
- Exceptional fall color

Resistance to insects and diseases

Drought tolerance

Tolerance of drainage problems

pH requirements

Salt tolerance

Light requirements

Known problems
- Pests
- Poor structure/weak wood
- Propensity to form surface roots
- Messy flowers/fruit/leaves
- Thorns

Maintenance requirements

Figure 6.3 Do not plant large-growing trees under utility wires.

Figure 6.4 The root system of this tree has outgrown its site and is now presenting a problem.

Figure 6.5 Parking-lot islands can be tough sites for trees because of high reflected light, low water, and sometimes de-icing salts.

drainage. Trees planted in a site that is too wet or too dry will often die within the first year. If the soil is compacted, tree growth may be greatly reduced due to insufficient oxygen in the root zone. Tree managers must be alert to soil contaminants such as road salt or herbicides that could poison the planting site (Figure 6.6).

Figure 6.6 Urban planting sites present harsh growing conditions where few trees will survive long-term.

Once the site has been analyzed, the next step is to consider the **design criteria**. Design criteria are based on the functions the tree is expected to serve. **Landscape functions** include engineering and architectural considerations such as controlling pedestrian traffic or hiding unsightly building features. Sometimes climate controls, such as wind screens or shade for an east or west exposure, are desired. Even aesthetic considerations such as colored foliage or flowers can be a design criterion. An ornamental tree may be planted as the focal point of a landscape (Figure 6.7).

Urban sites, especially street tree plantings, can pose some of the biggest challenges. Usually the underground space is limiting, and the soils are far from native. Soil compaction may be extreme, the pH may be very high, and moisture levels are likely to be suboptimal. Add to this the stresses of high reflected heat and light, de-icing salt injury, and the threat of vandalism, and it becomes clear why municipal arborists struggle to find trees that will survive. Selecting trees that can also meet the design criteria may be next to impossible. Some urban sites are simply not suited for trees.

Tree Considerations

Many factors must be considered when selecting a tree for a given site. These include cold hardiness, heat tolerance, **growth rate**, size at maturity, form, **pest resistance**, and maintenance requirements.

Specific plant characteristics may make a tree more desirable. These characteristics might include exfoliating bark, attractiveness to birds, or an interesting branching habit. Some trees are admired for their flowers or color change in the fall. Other plant characteristics, such as excessive leaf, fruit, or twig drop, may preclude the use of certain trees adjacent

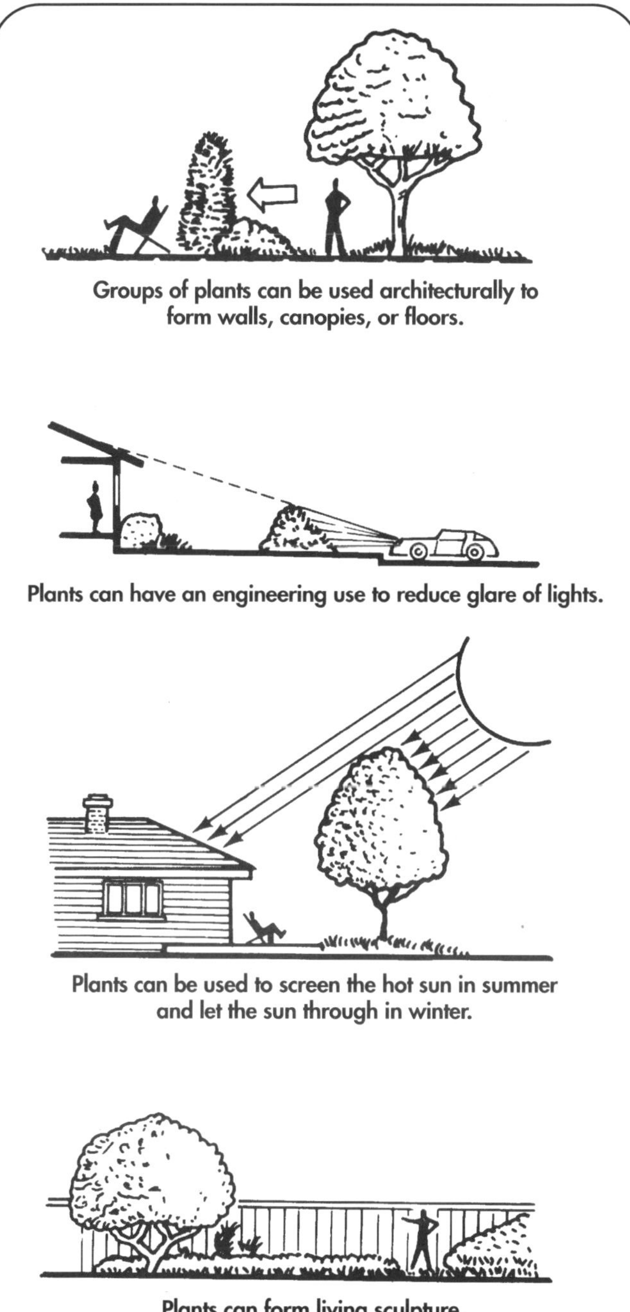

Figure 6.7 Functional uses of plants.

Figure 6.8 Some trees have very showy flowers.

Figure 6.9 Some trees have fruit that is ornamental and/or attractive to wildlife.

Figure 6.10 Other fruit can be messy or even a safety hazard.

Figure 6.11 Thorns can be a nuisance.

to a sidewalk, patio, or parking lot (Figures 6.8 through 6.11). Arborists should also consider future tree and site management when selecting tree species. Some plants require more water than others, and supplemental irrigation or drainage may be necessary in some landscape situations. Ensuring that adequate water is available is especially important in the first growing season following planting. Certain trees need to be pruned regularly. Avoiding plants that will require constant pest management programs is wise.

Figure 6.12 Cultivars can be selected for their growth forms. This upright form (left) is more appropriate for planting in restricted spaces.

Rapidly growing trees are often tolerant of poor soil conditions and owner neglect. Furthermore, they quickly reach a size where they provide certain functional uses such as shade and screening. However, rapidly growing trees often have weak, brittle wood that may break easily in storms and that decays readily. They usually do not live as long as trees with a more moderate rate of growth.

Selecting a tree with a particular growth form can sometimes offer a solution to a growing space problem. For example, a large tree with an upright (fastigiate) growth form and narrow branching habit could be planted closer to a building than one with a spreading form (Figure 6.12). Cultivars of some tree species are available with different growth forms such as upright, pyramidal, or weeping (Figure 6.13). Trees with different growth forms also provide a variety of architectural effects in the landscape.

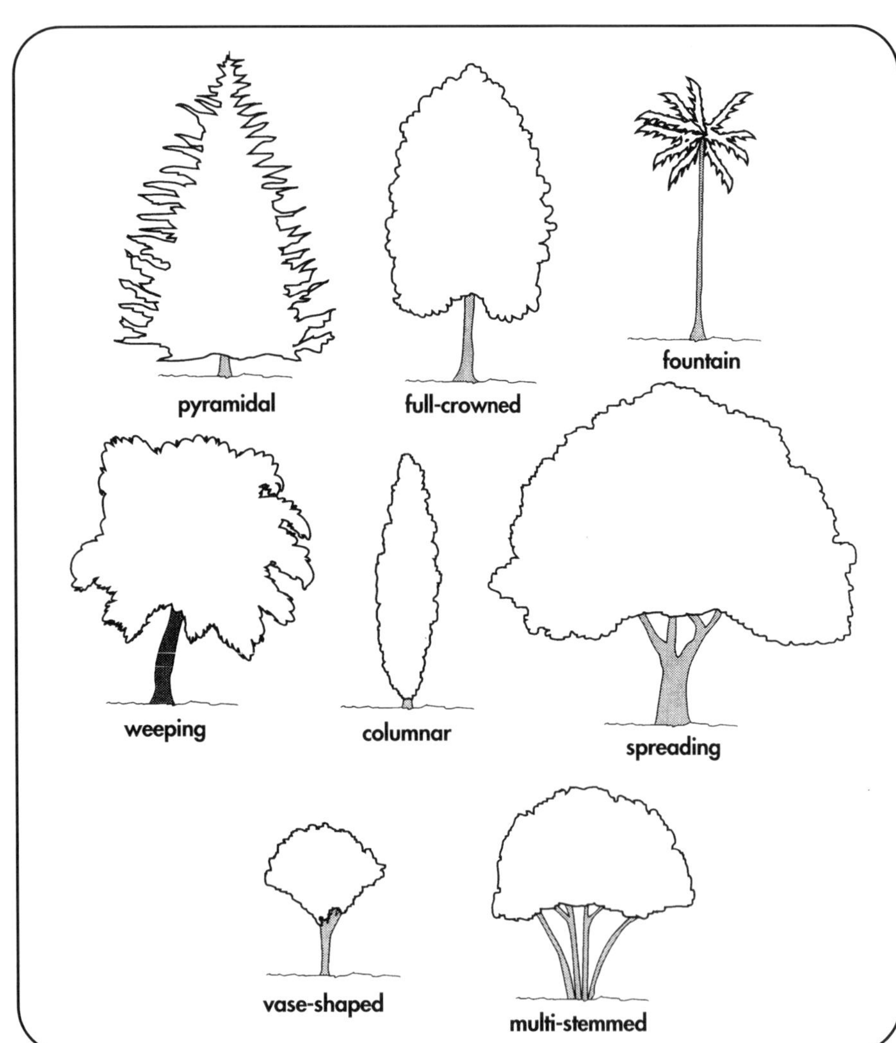

Figure 6.13 Growth forms of trees.

Too often we forget to consider the root system of a tree. The rooting space available can restrict the choice of trees that can be planted. Also, some trees have a strong tendency to form surface roots that can damage pavement and cause problems for lawn mowers.

Hardiness is a plant's ability to survive low temperatures. Many trees native to warm climates are not hardy in the north. The U.S. Department of Agriculture publishes a map of hardiness zones that labels each geographic area for low temperature extremes (Figure 6.14). Plant reference books and software list either the lowest (coldest) zone in which a tree will thrive or a range of zones where

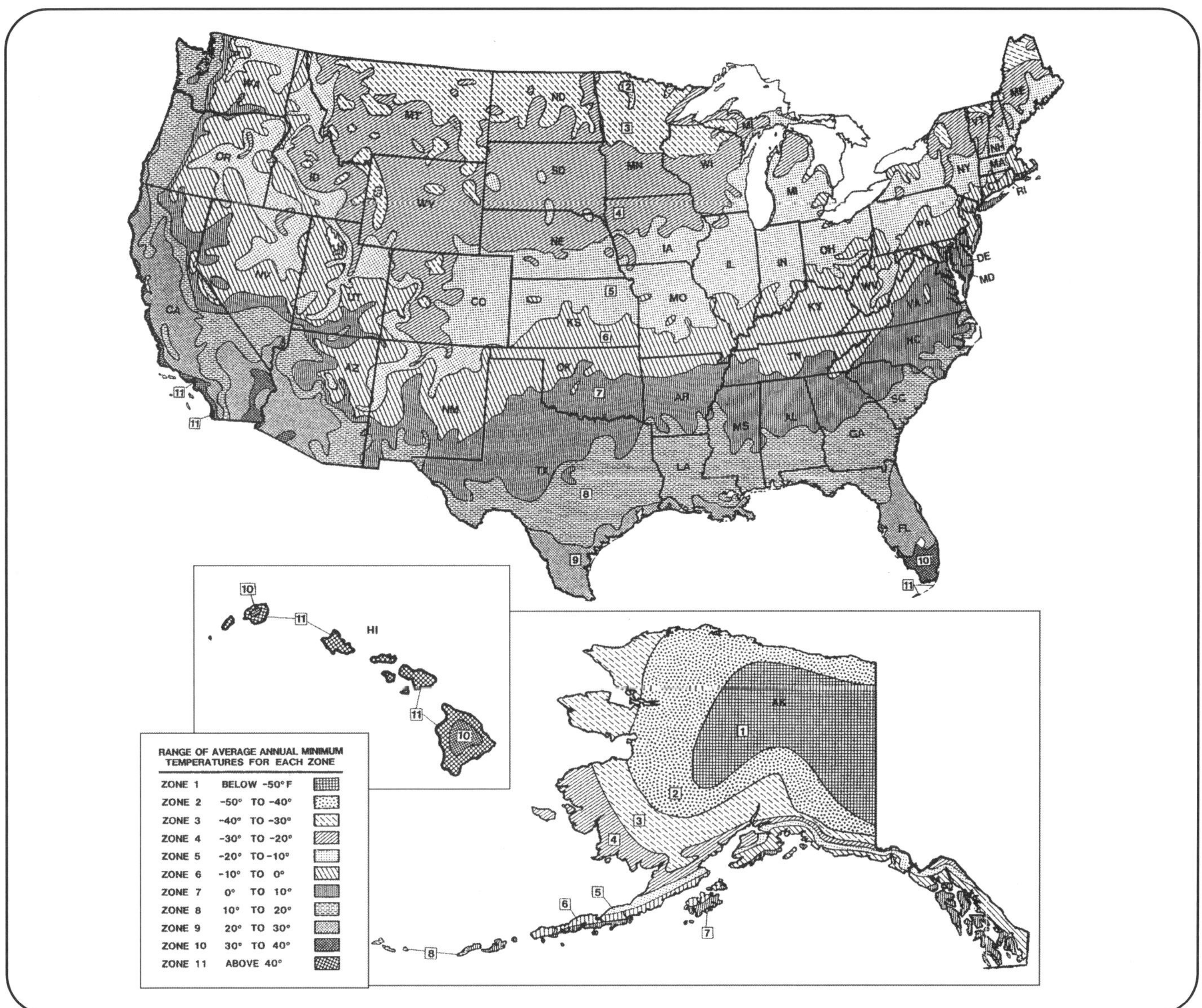

Figure 6.14 USDA hardiness zone map.

the plant will grow (e.g., zones 5 to 8). However, plants that are considered hardy in a given zone may still decline or die due to low temperatures if their roots are aboveground in containers or if the microclimate around the plant is unusually cold.

Just as some plants are not suited for severely cold temperatures, others may be stressed by high temperatures or dry conditions. For example, several of the white-barked birches are native throughout the northern and eastern United States. But if planted too far south, these trees are stressed by summer heat and are subject to borer invasion. Sometimes deodar cedar *(Cedrus deodara)* will grow well in California in warm zones but will do poorly in Florida and the Caribbean in the same hardiness zones. Some researchers are now working with heat zone maps that are similar in use to hardiness zone maps. With arborists taking a more holistic approach to tree health, more consideration is given to tree selection. Of particular concern is resistance or susceptibility to pests such as insects and diseases. It is important to select a tree that will not be stressed on a given site because stress can weaken a tree's ability to withstand the invasion of certain pests. Cultivars selected for their resistance to certain pests are occasionally available. For example, many crabapple (*Malus* spp.) cultivars are resistant to apple scab, a disfiguring disease of crabapples in temperate regions.

There has been considerable debate about the use of **native species** versus **introduced** (non-native) **species**. Native species, not surprisingly, usually grow well in their natural zones. For several hundred years, botanists and horticulturists have been introducing species of trees and other plants from different parts of the world. Some introduced species have been very successful, and some have become **naturalized species**, reproducing and thriving in their new settings for decades or even centuries. There are other instances, however, where introduced species have created problems. Some have been very invasive, virtually taking over landscapes. Other introduced trees have succumbed to pest problems for which there are no natural predators or parasites, or worse, introduced those pests to native species.

The most important thing is for arborists to select plants that will do well in a given site. In areas where new plantings are limited to native species by legislation, arborists sometimes have difficulty finding nurseries that grow appropriate species for urban plantings. Urban sites typically have altered soils and microclimates that do not mimic the native conditions of any tree. In addition, urban plantings must hold up to the cultural and maintenance practices of the site, as well as factors involved with human interaction. Professionals should choose from a list of plants that grow in a site most like the planting site. For example, if the site has compacted soil, trees that grow well in low-oxygen environments should be considered.

SELECTING TREES AT THE NURSERY

Once tree species have been chosen, the next step is selecting high-quality trees from the nursery. Most reputable nurseries produce nursery stock that meets the ANSI ASC Z60, *American Standards for Nursery Stock.*

For a tree to be a success when transplanted into a landscape, it is important to begin with a healthy plant. Nursery-grown plants are usually superior to trees collected from the woods or from unmanaged fields because more of the roots are contained in the root ball.

Look for a vigorous plant with good twig extension growth. Trees with good branch spacing and trunk taper are more desirable than those that have been headed back. Foliage should be evenly distributed on the upper two-thirds of the tree and not concentrated at the top. Avoid trees with many upright branches. Except for small-growing, multi-stemmed ornamentals, select trees with a single trunk or leader and with spreading branches. Check the plant for mechanical damage. Do not purchase a tree that has an injury to the trunk.

Healthy plants establish quickly in the landscape. Plants in poor health attract pests and require more maintenance. Examine the leaves and shoots. Choose trees with an abundance of healthy, green leaves. Check for the presence of insects or disease.

Examine the root ball of the tree. Balled and burlapped plants should have a solid ball that has been kept moist and protected from drying. If the plant is in a container, check the root system. Roots that are brown or black indicate a health problem. Avoid trees with circling or kinked roots.

Selection of the particular tree from the nursery can be as important as selection of an appropriate tree species. Selecting a healthy plant with good form can help ensure that the tree will be an asset in the landscape.

Chapter 6 Workbook

1. Name five tree species that would **not** be appropriate for planting under utility wires.

   a.

   b.

   c.

   d.

   e.

2. ____________________ is the ability of a tree to withstand low temperature and winter stresses in a given site.
3. True/False—Although a tree may be considered hardy in a given area, it may decline or die if the roots are unprotected.
4. Name three site characteristics that must be considered in tree selection.

   a.

   b.

   c.

5. Upright, pyramidal, and weeping are three examples of tree _______________ _______________ that are important in selection.
6. If a particular disease is known to be a problem, a tree species or cultivar should be selected that is _________________ to that disease.
7. Name three plant characteristics that may make a tree aesthetically desirable.

   a.

   b.

   c.

8. ____________________ is the process in which a tree adapts to its environment.
9. True/False—A tree listed as adaptable to wet soil conditions will always thrive if planted in those conditions.
10. Name five characteristics to look for when selecting a tree in the nursery.

    a.

    b.

    c.

    d.

    e.

CHALLENGE QUESTIONS

1. Give examples of the architectural and engineering functions of trees in the landscape.

2. Name five tree species adaptable to each of the following site conditions: wet soils, shade, high light, dry soil.

3. Name five plant attributes that may make a tree desirable for a given area.

4. Why is seed source or parent plant source so important in the adaptability of a tree to its site?

5. Discuss the merits and limitations of using only native species in new plantings as opposed to naturalized or introduced species.

SAMPLE TEST QUESTIONS

1. The climatic factor that determines hardiness zones is
   a. north–south location
   b. temperature, rainfall, and winds
   c. east–west location
   d. low temperature extremes

2. Trees to be planted under utility lines should be
   a. tolerant of heavy top pruning
   b. low-growing to remain below the lines
   c. vase-shaped or overarching to clear conductors
   d. all of the above

3. Some trees acclimate to shade conditions by
   a. developing larger leaves
   b. developing thinner leaves
   c. variegated foliage losing variegation, or colored foliage tending to be greener
   d. any or all of the above

4. Fastigiate trees have a growth form that is
   a. upright
   b. weeping
   c. overarching
   d. vase-shaped

5. Which of the following is a true statement?
   a. Lowland tree species will always grow well in wet soils.
   b. Forest understory plants tend to make good street trees because they don't grow tall.
   c. Some tree species are adapted to hot, dry, or bright light conditions with small, thick foliage and sunken stomata.
   d. Most evergreen conifers are very shade tolerant and tend to scorch in full sunlight.

Other Sources of Information

(See pages v–vi for complete bibliographic information.)

ANSI ASC Z60. *American Standards for Nursery Stock.*
Dirr, 1998. *Manual of Woody Landscape Plants.*
Gilman, 1997. *Trees for Urban and Suburban Landscapes.*
Gilman, 1998. *Horticopia: Trees, Shrubs, and Groundcovers.*
Sunset Books, 1995. *Western Garden Guide.*

CHAPTER 7

INSTALLATION AND ESTABLISHMENT

CHAPTER 7 INSTALLATION AND ESTABLISHMENT

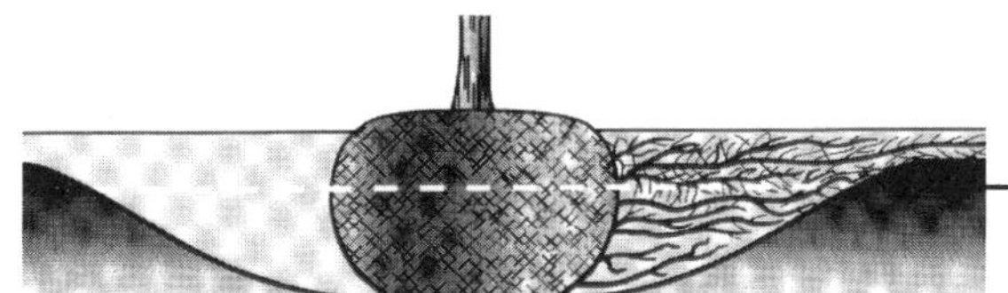

CHAPTER 7 INSTALLATION AND ESTABLISHMENT

Objectives

1. Learn the techniques and procedures used to plant and transplant trees. Understand how using proper techniques can improve survival chances and accelerate establishment.
2. Become familiar with the various ways trees are available from nurseries, the proper techniques for planting each, and the advantages and limitations associated with them.
3. Know the advantages and disadvantages of staking or guying newly planted trees. Learn the techniques used, and know when staking is appropriate.
4. Understand the special requirements of newly transplanted trees, and become familiar with early care procedures.

Key Terms

- balled and burlapped (B & B)
- bare root
- container grown
- containerized
- drum laced
- girdling root
- guying
- hardened off
- perched water table
- planting specifications
- root ball
- root pruning
- staking
- transplant shock
- tree spade
- tree wrap
- wire basket

INTRODUCTION

The key to giving a tree a healthy start in the planting site is using good planting procedures. Stress and physiological disorders can often be traced to poor planting practices. Many tree planting traditions have been passed down through generations in the landscape and arboriculture industry. While some of these techniques are still recommended today, others have been changed to reflect current research and technology. A well-informed arborist should be aware of the latest techniques in tree planting and transplanting.

PLANTING

A point that cannot be overemphasized is the importance of matching the tree and its requirements to planting site conditions. The best planting procedures known will not save a tree poorly suited for its site. The tree must be able to tolerate site conditions such as wet or dry soils, size limitations, or shade. Selecting a tree that meets site requirements is the single most important factor influencing the success of the tree.

Select a tree that is healthy and vigorous. The condition of the tree, particularly roots in the **root ball**, affects the chances for transplant success. New root growth depends on stored energy reserves inside the plant. Inspect the roots and aerial portions of the tree before planting. Roots should be abundant and light in color. Soft, brown or black nonwoody roots indicate a health problem. For balled and burlapped trees, check to see that the ball is solid, with little or no movement at the trunk. Wounds on the trunk or branches of the tree may become entry sites for insects or disease spores.

Trees are generally available from the nursery in one of three forms: **bare root**, **balled and burlapped (B & B)**, or **containerized**. There are other variations based on nursery production methods, such as pot-in-pot, temporary containers, and in-ground containers. Each form has its advantages and disadvantages. Site requirements or **planting specifications** often dictate the form to be planted. It is also important to select high-quality plants with good branching structure as described in the pruning chapter of this text.

Bare-root trees are usually small and easy to transplant. Because there is no soil on the root system, these trees are lightweight. It is vital that the roots be kept moist. Bare-root trees are normally planted during the dormant season before roots and buds begin to grow. If not planted immediately, bare-root trees should be stored cold (32°F to 40°F), with moist packing (or hydrogel dip) around the roots. Usually only deciduous trees and small conifer seedlings are handled bare root.

Bare-root trees should be planted on small, compacted mounds within the planting holes (Figures 7.1 and 7.2). The roots should be spread and distributed over the mound. Exposure of the roots to air must be minimized to reduce drying. Because the root system is limited, trees planted bare root may require staking, especially if there is a large canopy on a small root system.

Containerized trees' roots are contained in the soil or soilless production medium (bark, compost, peat, etc.) within the container. When selecting containerized trees, it is important to check the root system. Not all trees in containers are **container grown**. Often, bare-root trees are potted in containers and sold in the nursery. If they have recently been potted, there may not be an established root system in the container.

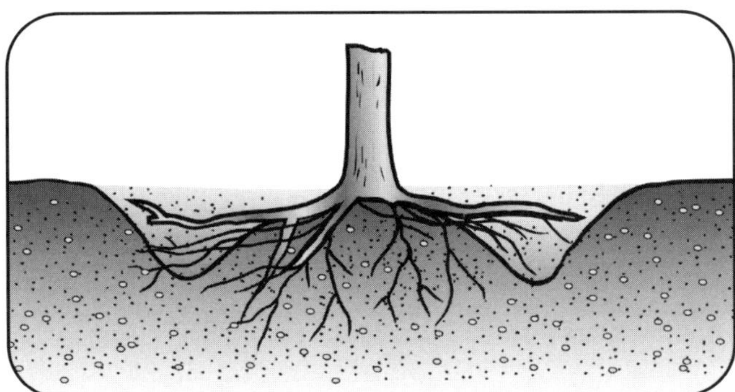

Figure 7.1 Planting bare-root trees. Spread the roots over the compacted soil mound.

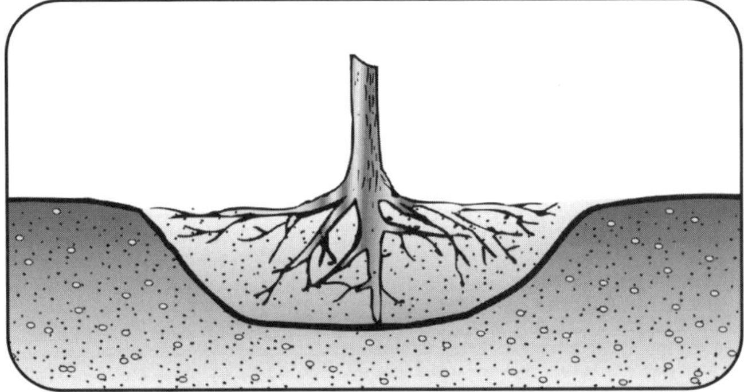

Figure 7.2 If there is a tap root present, plant the bare-root tree as shown. Staking may be required on bare-root trees.

Unless the container is biodegradable, such as a natural peat pot, it must be removed before planting. It may be preferable to remove the container anyway, unless the root system will not hold together. Sometimes roots will have grown in circles within the container. Circling roots should be separated and spread outward when planting. If roots are densely matted and cannot be teased apart, they should be cut in two places to stop the circling. Large roots that develop and grow across or around other roots at the root crown are called **girdling roots** (Figure 7.3). Girdling roots can choke off vascular tissues in the tree. This can lead to decline and death of limbs or even the entire tree.

If properly watered and maintained, container-grown trees can be planted any time of year. If planted in the fall after leaf drop, roots may begin establishment before dormancy. Early spring, before budbreak, is also a good time to plant most trees. Roots begin growing immediately, and light, temperature, and soil moisture levels are usually good for plant establishment. Trees establish most quickly when soil temperatures are warm and moisture availability is adequate. The most important factor in successfully planting containerized or container-grown trees is maintaining adequate soil moisture to encourage the roots to grow out into the surrounding soil.

Many trees are balled and burlapped in the nursery. Trees are dug with root balls intact, and wrapped with burlap. As much as 95 percent of the absorbing roots can be lost in digging, but some roots are preserved in the root ball. The burlap used to wrap the root ball supports the root ball and helps keep roots from drying out from exposure to air.

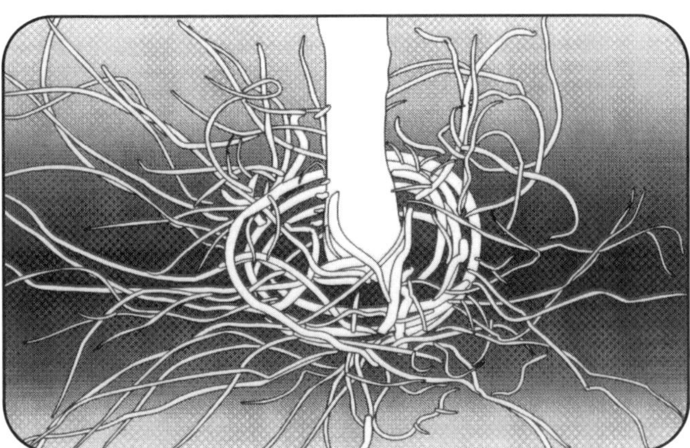

Figure 7.3 Girdling roots may form from roots that circle around when restricted in containers.

Natural fiber burlap is biodegradable, but some nurseries use treated or synthetic burlap. It is important to remove the burlap if it is synthetic or treated. It is also advisable to fold back natural burlap to decrease the "wicking" effect, enhance water infiltration (burlap can repel water), and ensure that roots can grow out of the root ball (Figure 7.4). Often the burlap is held in place with twine. All twine or ropes tied around the trunk of the tree must be removed to avoid girdling the tree. Remove all tags and labels to prevent them from girdling the trunk or branches as the tree grows.

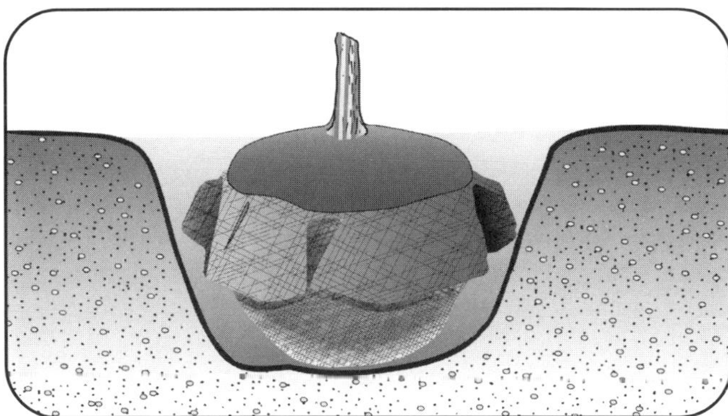

Figure 7.4 Cut all binding twine, and cut or fold the burlap back from the top of the ball to allow easy water movement into the root ball.

Some larger balled and burlapped trees come in **wire baskets** to maintain the integrity of the ball during handling. Baskets can sometimes last decades in the soil, and they can partially girdle roots, restricting vascular transport. Although it may be impractical to remove the entire basket, it is preferable to cut away as much of the wire as possible, once the tree is in the planting pit and the ball is stabilized (Figure 7.5). Wire removal eliminates interference with rakes or lawn mowers and allows roots to grow and spread freely near the surface.

Once the tree has been selected, the planting hole can be dug. Ideally, the planting hole for a tree should be two to three times the width of the root ball at the soil surface, sloping down to about the width of the root ball at the base. Most new root growth will be shallow and horizontal (Figure 7.6). Root growth from the bottom half of the ball is often reduced due to inadequate drainage and aeration. On poorly drained sites, roots in the lower half of the root ball may die. The most vigorous root growth occurs near the surface.

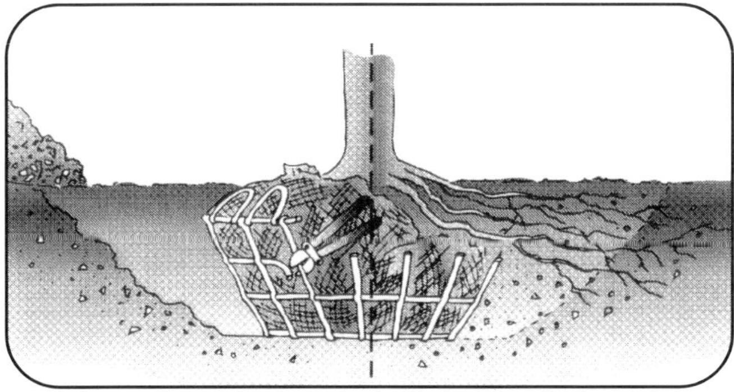

Figure 7.5 Cut away the upper one-third to one-half of the wire basket to avoid future problems.

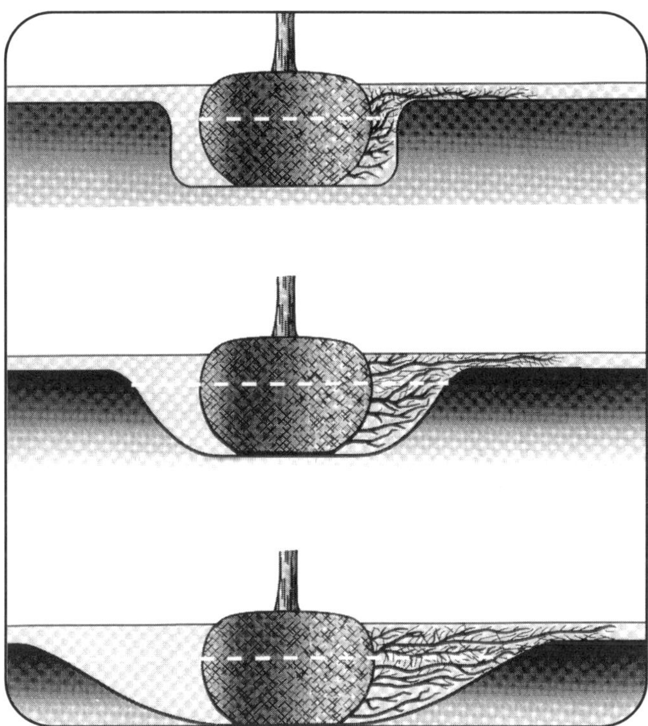

Figure 7.6 Root growth from the root ball varies with the shape of the planting hole. The broad planting pit (bottom drawing) provides for optimal new root growth.

The hole should never be deeper than the root ball. The planting hole may act as a dish and hold water, especially in clay soils. Oxygen levels are low in the bottom of such holes, and they are not conducive to rapid root growth (Figure 7.7). One of the most common planting problems is planting too deeply. Deep planting can even be a problem when professionals plant trees because containerized and balled-and-burlapped trees often arrive from the nursery with soil too high up the trunk due to production techniques. It is imperative that the natural root flare be located before planting. The top of the root ball should be even with or slightly higher than soil grade at planting. Planting a tree too deeply can stress the tree and drown or suffocate roots.

Soft fill should not be added to the bottom of the hole because the root ball will settle and be planted too deeply. In areas where the soil is dense clay, the

Figure 7.7 If water does not drain readily from the planting hole, excess water may drown the roots of a tree planted in the hole.

tree should be planted slightly shallow, with 3 to 5 inches of the ball higher than the original grade. An alternative is to amend an area far larger than the planting hole with an organic material such as compost. Soil should not be placed over the top, but should be tapered out from the ball to the original grade. The exposed ball should be covered with 2 to 4 inches of mulch, but not up against the trunk.

Adequate drainage is an important consideration in successful planting. Poor drainage—characteristic of heavy, clay soils—accounts for many planting losses. When planting large trees in poorly drained soils, install a tile drainage system. For smaller trees, planting shallow may be sufficient to allow roots to develop in better-aerated soil. Do not put gravel in the bottom of the planting hole; it does not aid drainage. Water will accumulate in the finer-textured soil above the coarse gravel layer until the soil is completely saturated. This is known as a **perched water table**.

The tree should never be lifted by the stem, except bare-root trees. The root ball must be handled carefully. Small, fibrous absorbing roots are easily broken. Place the tree in the planting hole gently. Check to see that the top of the root ball will be no deeper than the soil grade. If practical, the tree should be oriented in the hole so that it faces the same direction as it did when it was dug. In certain climates, thin-barked species, such as maple (*Acer*) and beech (*Fagus*), are susceptible to sunscald.

In most cases it is best to backfill the hole with the same soil dug out of the hole. Research shows that soil amendments do not assist in tree establishment and growth unless the natural soil is so poor that it significantly restricts root growth. If the natural soil is extremely poor, it may have to be amended to improve structure, water-holding capacity, or drainage. However, the backfill soil type should match the soil type of the site as closely as possible. Backfilling with a sandy loam in heavy clay may cause the planting hole to collect water and suffocate the roots. If the backfill must be amended, the hole should be very wide to allow for several years' growth within the new soil.

Work the soil around the ball so that no air pockets are left. Large pockets of air can allow the roots to dry out. Firm the soil around the bottom of the root ball so that the tree is vertical and adequately supported. Unless drainage and excess moisture are problems, the backfill can be watered in when half filled. The remainder of the backfill need only be tamped slightly as the hole is filled. Water thoroughly and slowly after backfilling. The remaining soil is sometimes mounded into a dike or berm beyond the outer edge of the root ball to collect water over the root zone, especially on sloped sites (Figure 7.8).

Figure 7.8 Soil is sometimes mounded into a dike or berm beyond the outer edge of the root ball to collect water over the root zone, especially on sloped sites.

TRANSPLANTING

Transplanting a tree involves the additional procedures of digging and preparation for moving. Some species transplant more easily than others. Because digging a tree for transplanting can remove as much as 95 percent of the absorbing roots, trees that are difficult to transplant should be moved when conditions are optimal. Some trees can even be moved in the summer if extra care is taken. If the trees are dug from a nursery field, they may be stored in a protected holding area to be **hardened off.** New root generation begins and some leaves may drop due to reduced water availability, which can reduce transpiration.

In general, the best time to transplant most tree species is in the early spring or fall. Many deciduous plants can be moved just after leaf drop in the fall. The moisture level in the soil is relatively high and the soil is still warm. Roots of some species have a chance to grow and begin to establish before the ground freezes. Some plants are more easily transplanted in the spring before budbreak. Transplanting dormant trees reduces demand on soil moisture because transpiration is minimal. Evergreen trees are also more easily transplanted while dormant. Sometimes, very large trees are moved in the winter when the ground is frozen (Figure 7.9). Heavy equipment can work around the tree while doing minimal damage to the surrounding area. If the outer 4 to 6 inches of the root ball are frozen, it will be easier to move without damage. Frozen root balls require less wrapping, are more cohesive, and are easier to handle. Sometimes the tree can be pre-dug and thoroughly mulched before the ground freezes. The new hole should also be prepared before freezing.

Figure 7.9 This photo shows a large tree being moved in the winter (Chas. F. Irish Co. 1940).

Techniques used in digging vary little with species, size, and soil type. The procedures described here are for digging and transplanting a large tree. The techniques are similar for smaller trees that are easier to move.

If the transplanting is planned well in advance, it is possible to improve the chances of survival and decrease the establishment period by **root pruning**. The idea is to dig a small trench around the tree at a radius smaller than the radius of the final root ball. The tree then produces many new roots from these cuts, producing a more densely rooted ball. Sometimes this procedure is done more than once to maximize root density. The second cut would be out farther than the first, and the final ball would be out farther yet. Root pruning should take place early enough to allow time for new root growth but not so early as to allow the new roots to spread widely.

Before moving, if necessary, the lower branches of the tree should be tied to prevent injury or breaking. Care must be taken to avoid damage to the bark. Do not tie the limbs so tightly that a sharp bend is created that could compress the tissues or break the limbs.

The tree should be measured 12 inches above the root crown to determine diameter. A rule of thumb for the width of the root ball is a minimum of 10 inches in diameter per inch of trunk diameter. Thus an 8-inch tree should have a root ball at least 80 inches in diameter. The depth of the root ball may vary with the root growth of the tree. Root growth is deeper in certain species than in others, and a drier soil tends to cause deeper rooting. In general, a root ball depth of 30 to 36 inches is sufficient. It can be less for smaller trees or trees that are not rooted deeply.

The first cuts around the perimeter of the root ball should be made with a sharp spade or Pulaski ax. Cuts must be clean to avoid tearing or breaking the roots. If the ball is being dug with machinery such as a backhoe or trencher, the initial ball should be dug several inches larger than the final ball size. The shaping and final cuts should then be done by hand. As larger roots are located, they can be cut with loppers or a saw. The roots should be cut both at the edge of the ball and the outer edge of the digging trench to clear the trench for more digging. While digging the trench, avoid standing on the root ball. The edge of the ball could break down and damage the roots.

Once the ball has been dug to the desired depth, it can be shaped. The ball should taper on the sides, slanting inward toward the base (Figure 7.10). A root ball with a 9-foot diameter at the top might measure 6 to 7 feet at the base. The ball should stand on a pedestal of soil for shaping and burlapping before it is undercut.

Figure 7.10 The ball should be shaped before burlapping, taking care to make clean cuts on the roots.

Burlap is placed on the sides and across the top of the ball and is pinned together using nails. Tucks may be taken to pull the burlap snug. The burlap should cover the full circumference of the root ball, with a bottom skirt of burlap hanging over the pedestal. This skirt is pinned to the ball later, after the tree is taken out of the hole.

For additional support, large soil balls should be **drum laced** with rope. The rope generally used is ½-inch manila or double strands of 4-ply sisal. One rope is looped around the bottom of the ball and pulled snug at the top of the pedestal of soil. A second loop of rope is placed on the top of the ball and may be held in place with nails. One end of the lacing rope is tied to the top loop and laced up and down the circumference of the ball, weaving in and out of the upper and lower loops.

The bottom rope is then pulled tight. The top rope is snugged up and securely tied. The slack is worked out of the side lacing, working around the ball until each section is tight. If the soil ball is large, top lacing should also be used. A top rope is laced back and forth across the top of the ball, with care being taken not to let the rope cut into the trunk of the tree.

Once the ball is laced and secured, the tree can be undercut. One relatively easy way to make a clean cut is to use a cable. Cable cutting can be accomplished without having to tilt the ball of soil. An anchor pin is driven into the bottom of the hole opposite a trench where the cable is to be pulled. The cut-off cable is a ½-inch or ⅝-inch cable with an eye in each end. The cable must be long enough to pass around the ball and out to where the winch is hooked to one eye. The other eye is placed over the steel anchor pin. The pulling end must pass under the anchor end to keep the cable from riding up into the ball. The cable is pulled using a winch, and the ball is undercut.

Large trees must be removed from the hole with a crane or other mechanical device. Chains or large slings are placed around the ball and attached to the crane hook. Trees should not be lifted by the trunk because this can cause trunk injury and serious damage to the root ball. Once the tree is out of the hole, the burlap should be fastened to the bottom of the ball (Figure 7.11).

If the tree must be transported to a distant site, it should be protected. The trunk should be well padded to protect from injury. The crown of the tree should be loosely wrapped with a tarp or burlap to minimize drying and wind damage. Antitranspirants may be applied but should not be considered a substitute for adequate watering. Special permits may be necessary to transport large trees on public roads.

Figure 7.11 Carefully wrap, lace, and protect the tree before attempting to lift the ball.

Digging and Planting with a Mechanical Tree Spade

A **tree spade** is a mechanical device used to transplant trees (Figure 7.12). The spade encircles a tree and forces several large blades diagonally into the ground to form a root ball. The blades can make clean cuts on the roots unless the roots get caught between blades. Mangled roots can be severed cleanly after the tree is removed from the hole. The tree is lifted hydraulically from the hole, and it can be transported vertically or tilted.

Tree spades are available in various sizes. No attempt should be made to transplant trees larger than the size limitations of the spade. Although larger trees can be removed from the hole, transplant survival is not likely because the size of the ball must be proportional to the tree size.

Tree spades are often used to dig planting holes. However, holes dug with tree spades (as well as

Figure 7.12 Tree spades can be used to dig and transplant trees. If the planting hole is also dug with a tree spade, it should be enlarged and the sides roughed up.

augers) tend to be glazed on the sides, which can inhibit root penetration. If tree-spade–dug trees are planted in holes dug with the same spade, there will be little area for new root development within the hole. If tree spades or augers are used to dig the hole, the walls should be roughened and widened prior to installation. One good technique is to place the tree in the new planting site and then hand-dig 1 to 2 feet wide around the hole, working the soil down around the root ball. This method loosens the soil around the ball and helps eliminate any gaps that could allow roots to dry out.

If tree spades are used for transplanting, care must be taken to obtain a vertical root ball and to plant the tree upright. Planting on a slope can be a problem because the ground level will be different. Remember that the tree must be planted vertically in the new site.

STAKING AND GUYING

Staking of newly planted trees is not always necessary. In fact, staking can have detrimental effects on the development of a tree. When compared to trees that have not been staked, staked trees produce smaller caliper and less trunk taper, develop a smaller root system, and are more subject to breaking or tipping after stakes are removed. In addition, staked trees may become injured or girdled from the staking or guying materials.

Some trees, however, cannot stand upright without support. Bare-root trees, trees grown in small containers with a loose potting mix, and large conifers may require support while they establish, especially in windy sites. In some instances, staking may be recommended to reduce movement of the root ball and subsequent damage to the fine, absorbing roots. In urban sites, stakes are sometimes used to protect young trees from mechanical (equipment) damage and to reduce vandalism.

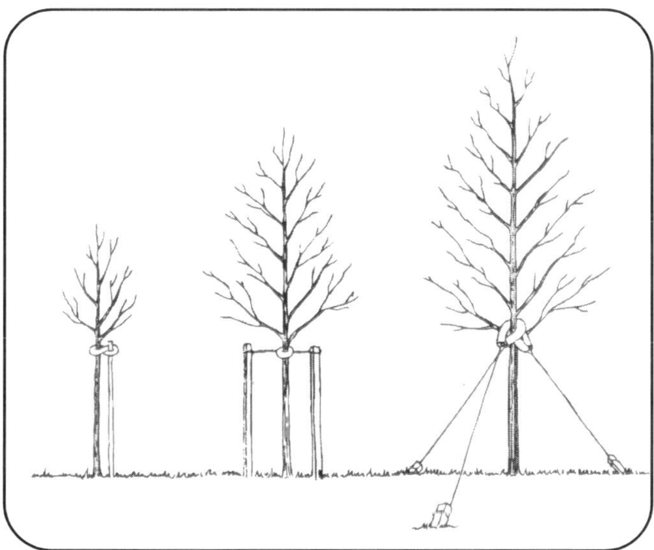

Figure 7.13 If a single stake is used, it should be placed on the upwind side of the tree. Stakes should be just high enough to keep the tree upright without the top bending above the tie point. Stakes for guy wires should be driven into the ground in the same direction as the guy wires.

Normally, one or two stakes are used to support a tree (Figure 7.13). If a single stake is used, it should be placed on the upwind side of the tree. The stake should not be driven through the root ball because this could damage the roots. The material used to attach the tree to the stake should be broad, smooth, and somewhat elastic. To allow for flexibility, the tree should be tied with a figure-8 loop or proprietary tie between the tree and the stake (Figure 7.14). Stake the tree as low as practical along the trunk while still providing the necessary support. If the tree is tied too rigidly to the stake, the tree will develop a less sturdy trunk and root system and may be more subject to girdling and breakage above the tie.

If two support stakes are used, a single, flexible tie attached to the tops of the stakes will be sufficient for support. The stakes should be high enough to keep the tree upright without the top bending above the tie point. Some municipal arborists specify that trees planted in urban sites be staked with three stakes. Triple staking provides more protection against wind, lawn mowers, and vandals.

Another method of supporting larger trees is to anchor the root ball with stakes or belowground guy wires (Figure 7.15). Trees greater than 4 inches in diameter are often supported by **guying**. Trees are generally guyed with three or four wires anchored in the ground. Anchoring devices include stakes, land anchors, and deadmen. If stakes are used, they should be driven in line with the guy wire pointing toward the tree. Stakes driven perpendicular to the guy wire tend to loosen.

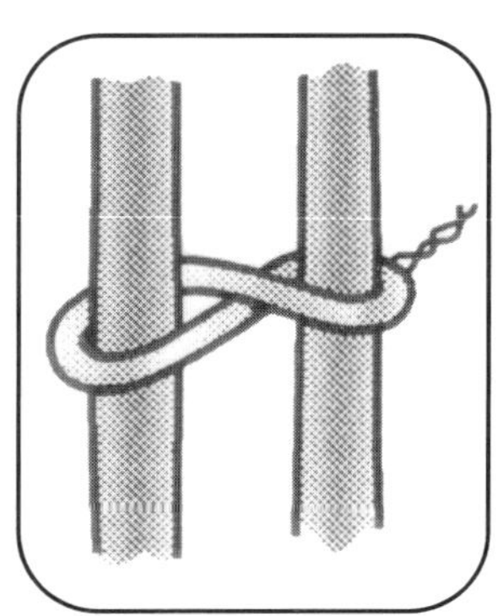

Figure 7.14 Use a broad, flexible material to attach a tree to the stakes.

A common technique passes guy wires through a section of hose to protect the tree. The wires and hose are passed around the tree at crotches, and the wires are twisted to tie them off. Caution is in order when using this technique because the tree can be partially girdled despite the limited protection of the hose. Some of the new materials on

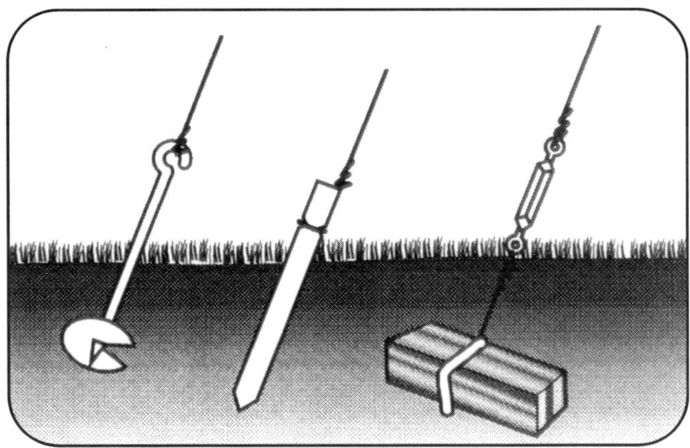

Figure 7.15 Various anchoring devices used for staking trees.

the market for staking trees can reduce this problem. Any modification to reduce girdling and friction will minimize damage to the tree.

On larger trees, guy wires can be attached to the tree with eye screws. This procedure does wound the tree but causes minimal damage to a vigorous tree and eliminates the chance of girdling by guy wires. The size of the eye screw or eyebolt used should be proportionate to the size of the tree. If the eye screws cannot be removed readily when the guy wires are removed, they should be left in place.

Staking or guying systems should be checked within one year to be sure they are not injuring the tree. Support stakes or guy wires should generally be removed after one growing season. If support systems are left in place for more than two years, the tree's ability to stand alone may be reduced, and the chances of girdling injury are increased.

CARE AFTER PLANTING

Transplanting is a major operation from which most trees recover slowly. With field-grown (B & B) trees, a major portion of the root system is lost in digging, and the tree must re-establish sufficient roots to sustain itself. The tree's ability to absorb and transport water and minerals is greatly reduced. Varying degrees of water stress are the result, and the tree experiences post-planting stress referred to as **transplant shock**. This is less of an issue with container-grown trees.

The rate of recovery and re-establishment after planting or transplanting varies with the species, planting season, soil, site conditions, moisture availability, and climate. The size and vigor of the plant can play a major role as well. The general rule of thumb for re-establishment in temperate climates is one year for each inch of tree caliper (diameter). Because of the time difference for recovery, it is sometimes possible for smaller-diameter transplants to recover more quickly and outgrow larger transplants.

Because the root system of a newly planted tree is limited, fertilization is often not recommended at the time of planting. Excessive fertilizer salts in the root zone can be damaging. Phosphorus fertilizer is often thought to stimulate root growth. There is no evidence of this unless phosphorus is deficient in the soil, which is rarely the case. If fertilizer is used in the first growing season, application of a slow-release fertilizer to provide a low rate of nitrogen is suggested.

Pruning following planting should be limited. There is no advantage to removing one-third of the crown. In fact, a tree will grow and establish most rapidly if pruning is minimized at planting. Diseased, dying, broken, and damaged limbs should be removed. The spacing, balance, and attachment of the limbs should be evaluated. Some limbs may be removed for structural stability or appearance. The central leader should not be removed. Proper training of young trees to produce a strong branch structure is important, and the techniques involved are discussed in the pruning chapter of this text. Generally, however, it is preferable to start these procedures after the tree has become established.

Many early references recommend wrapping the trunks of newly planted trees to protect against temperature extremes, sunscald, boring insects, and drying. More recent research indicates that temperature differentials at the bark are greater with **tree wrap** than without. Further, tree wrap tends to hold moisture on the bark and can lead to fungal problems. Also, insects tend to burrow between the bark and the wrap and can be worse with wrap than without it. Wraps (especially those made of white polypropylene) may be beneficial for late spring and summer when planting into hot sites such as parking lots, but they should be removed after one year.

Sometimes tree guards made of plastic or metal are installed around the trunks of newly planted trees. These guards have the advantage of minimizing physical damage caused by animal feeding, mowers, string trimmers, or cars. If guards are installed, they should be loose fitting and should allow air to circulate around the trunk.

The area around the tree should be mulched with about 2 to 4 inches of mulch. Organic mulches are usually preferred. The mulch helps reduce competition from weeds and grasses, conserve soil moisture, and moderate soil temperature extremes. Mulch should not be placed against the stem of the tree because that can cause bark suffocation or crown rot. The broader the mulched area, the better; but deeper is not better. "Volcanoes" of mulch around trees can restrict oxygen and water availability to the roots and can lead to root rot. Black plastic should not be placed under the mulch because it restricts water movement and oxygen availability to the roots.

Proper watering is the key to survival of newly planted trees. If rainfall is not sufficient (generally 1 inch per week), the tree should be watered every five to seven days. A slow, gentle soaking of the root zone is preferable. It is very important to monitor the soil moisture level of the root ball itself, especially in the first few weeks after planting. Often the soil or growing medium from the nursery will dry out more quickly than the surrounding soil. Also, the ball may actually repel water, especially if it is wrapped in burlap. If the soil around the root ball dries out too much, root growth will be halted and the establishment period will be lengthened. If it stays dry for too long, the tree will not survive. Although adequate water is critical to survival and new root growth, excess water accumulation in the planting hole is a leading cause of transplant death. Watering must be appropriate for soil type and drainage.

A comment about planting specifications is in order. Arborists often work with other green-industry professionals (landscape architects, designers, contractors) and must either write or work according to planting specifications. The details and scientific research bases of these specifications vary and, unfortunately, some do not take into account the latest techniques and procedures. Sample planting specifications are available from several green-industry organizations. One good example is included as an appendix to *Principles and Practice of Planting Trees and Shrubs* by Watson and Himelick.

Chapter 7 Workbook

1. Trees are generally available from the nursery in one of three forms.

   a.

   b.

   c.

2. Bare-root trees are normally planted when _______________, before buds begin to grow.
3. ________________ roots can become a problem because they can constrict the vascular system in the trunk or in other roots.
4. Planting holes should be dug __________ times the width of the root ball at the surface, with the sides sloping down to the diameter at the base of the root ball.
5. Trees that are dug in the nursery are wrapped with _______________ to help keep the root ball intact and reduce exposure of the roots to air.
6. The planting hole should never be ____________ than the root ball.
7. In areas where the soil is heavy clay, the tree should be planted slightly _______________.
8. Do not put _______________ in the bottom of the planting hole to improve drainage.
9. True/False—Research has shown that soil amendments generally do not assist the tree in establishment and growth.
10. True/False—Digging a tree for transplanting can remove as much as 95 percent of the absorbing roots.
11. The two best times to transplant most trees are ___________ ___________ and ____________.
12. When transplanting a tree, the root ball should be _____ inches in diameter for every inch of tree caliper. In general, a root ball depth of ____________ should be sufficient.
13. True/ False—If trees have wire baskets to help maintain the integrity of the root ball, these baskets should *never* be removed or cut down at planting.
14. True/False—Most of the new roots generated after planting will grow horizontally and near the soil surface.
15. Pre-digging to create a more densely rooted ball is called ______ ___________.
16. True/False—Staking of newly planted trees is not always necessary.
17. Name three adverse effects of staking or guying trees.

    a.

    b.

    c.

18. If a single stake is used to stake a tree, it should be placed on the ______________ side of the tree.
19. True/False—The material used to attach the tree to the stake should be broad, smooth, and flexible.
20. Warm soil temperatures and adequate soil ______________ are the optimal conditions for new root growth.
21. Transplant shock is mainly due to ____________ stress from the greatly reduced root system.
22. If fertilizer is applied at planting, it should be a __________ _______________ type to avoid excess salt buildup in the root zone.

23. True/False—There is no advantage to pruning one-third of the tree crown at the time of planting.
24. The most important maintenance factor in the survival of a newly planted tree is proper ________________.
25. True/False—Tree roots may suffocate if the tree receives too much water after planting.

CHALLENGE QUESTIONS

1. What are the advantages and disadvantages to planting each of the following ways: bare root, balled and burlapped, and containerized?

2. Why is the use of soil amendments in the backfill of planting holes no longer routinely recommended?

3. Why are most trees easier to transplant in the fall or early spring?

4. What are some of the physiological effects of staking on the early growth and establishment of a tree?

5. Pruning of one-third of a tree's crown to compensate for root loss when transplanting has not proved to be beneficial. Why might this be the case?

SAMPLE TEST QUESTIONS

1. Staking or guying when planting a tree
   a. is done only for bare-root trees
   b. is not necessary for trees greater than 6 inches in diameter
   c. is not always required or necessary
   d. promotes a larger and stronger root system and better trunk taper
2. When planting in a compacted, clay soil
   a. the backfill should be modified 50 percent with peat
   b. the tree should be planted slightly shallow (with the top of the ball a few inches above ground level)
   c. the hole should be dug 6 to 8 inches deeper than the ball and soft fill added
   d. the tree should be planted deeper to discourage surface root growth

3. Placing gravel in the bottom of the planting hole in a clay soil site
   a. will improve drainage
   b. will prevent the formation of girdling roots
   c. will restrict soil space and create a perched water table
   d. will improve aeration and water infiltration

4. When planting a container-grown tree,
   a. separate and tease apart the roots to reduce girdling root formation
   b. place soft fill in the bottom of the planting hole to encourage taproot growth
   c. backfill the hole with a soilless growth medium to encourage root growth
   d. none of the above

5. The most important reason to prune a tree when transplanting is
   a. to compensate for root loss
   b. to invigorate the tree
   c. to reduce growth at the tips
   d. to remove structurally weak or damaged branches

Other Sources of Information

(See pages v–vi for complete bibliographic information.)

ANSI ASC Z60. *American Standards for Nursery Stock.*
Gilman, 1997. *Trees for Urban and Suburban Landscapes.*
Harris et al., 1999. *Arboriculture: Integrated Management of Landscape Trees, Shrubs, and Vines.*
Watson and Himelick, 1997. *Principles and Practice of Planting Trees and Shrubs.*
Whitcomb, 1991. *Establishment and Maintenance of Landscape Plants.*

CHAPTER 8
PRUNING

CHAPTER 8 PRUNING

CHAPTER 8

PRUNING

CHAPTER 8 PRUNING

A
B
C
C
B
A
branch collar
branch collar

CHAPTER 8
PRUNING

Objectives

1. Know why, when, and how a tree should be pruned.
2. Understand how trees respond to pruning and the effects of severe pruning on a tree.
3. Understand the relationship of the branch collar and branch size to wound closure and the potential for decay.
4. Describe the procedures and techniques used in pruning. Become familiar with the terms used to describe pruning techniques.

Key Terms

ANSI A300 standards
antigibberellins
branch bark ridge
branch collar
branch protection zone
codominant
compartmentalization
crown cleaning
espalier
fronds
heading back
included bark
internodal
lateral
leader
lion tailing
permanent branches
plant growth regulators
pollarding
raising
reduction
restoration
scaffold branches
structural pruning
subordinate
temporary branches
thinning
topping
utility pruning
vista pruning
watersprouts
wound dressing

INTRODUCTION

Pruning is the most common tree maintenance procedure. Forest trees grow quite well with little or no pruning, but in landscape situations, tree pruning is often desirable or necessary to remove dead branches, improve tree structure, enhance vigor, or maintain safety. The fact is, most pruning is done for "people reasons." Trees growing in the landscape or urban forest must endure stresses not found in the forest, and human safety and aesthetic preferences dictate certain pruning requirements. Yet arborists must understand the biology of trees and their basic requirements in order to optimize the health and structure of trees through pruning.

Pruning cuts must be made with an understanding of how the tree will respond to the cut. Improper pruning can cause damage that remains for the life of the tree. This chapter is designed to help the arborist become aware of tree response to the various methods of pruning. Knowledge of the tree's response will help the arborist achieve the goal of a healthy, aesthetically pleasing tree.

Removing foliage from a tree by pruning branches affects its future growth. Removing leaves reduces the tree's overall photosynthetic capacity and may reduce overall growth, which can create a dwarfing effect. At the same time, growth after pruning takes place on fewer shoots, so unpruned parts tend to grow more than they would have without pruning. This is called shoot invigoration. These principles should be considered when pruning trees and other woody plants.

In the United States, pruning should be done in accordance with the American National Standards Institute **(ANSI) A300 standards.** Europe and Australia have standards as well.

REASONS FOR PRUNING

Because each cut has the potential to change the growth of a tree, no branch should be removed without a reason. Common pruning goals are to improve tree structure or health, or to accommodate human needs such as clearance pruning or hazard mitigation. Trees may also be pruned to increase light penetration or to provide a view. In most cases, tree pruning is of a corrective or preventive nature.

TIMING

The best time to prune trees depends on the desired results. As a rule, growth is maximized if pruning is done just before the buds swell, in early spring. Pruning when trees are dormant can minimize the risk of pest problems associated with wound entry and allow trees to take advantage of the full growing season to begin closing and compartmentalizing wounds. Plant growth can be reduced if pruning takes place during or soon after the initial growth flush, so pruning at this time is usually not recommended. This is when trees have just expended a great deal of stored energy to produce foliage and early shoot growth. Removal of many live branches at this time can stress the tree. A few tree diseases, such as oak wilt, can be spread when pruning wounds allow spores access into the tree. Susceptible trees should not be pruned during active transmission periods.

Flowering can be prevented or enhanced by pruning at the appropriate time of the year. Landscape trees that bloom on current season's growth, such as crape myrtle (*Lagerstroemia* spp.) or linden (*Tilia* spp.), are best pruned in winter, prior to leaf emergence, or in the summer after bloom has occurred. Plants that bloom on last season's wood, such as many fruit trees, should be pruned just after bloom. Often, fruit trees are pruned during the dormant season to enhance structure and distribute fruiting wood, and are pruned after bloom to thin fruit.

Some references recommend that certain species of trees, such as maples (*Acer* spp.) and birches (*Betula* spp.), not be pruned in the early spring when sap flow is heavy. These trees tend to "bleed," or drain sap from the pruning cuts. Although unattractive, research has shown that sap drainage has little negative effect on tree growth. Most routine pruning and removal of weak, diseased, undesirable, or dead limbs can be accomplished at any time with little negative effect on the tree.

PRUNING CUTS

Each cut should be made carefully, at the correct location, leaving a smooth surface with no jagged edges or torn bark. Pruning cuts may be classified by where they are made on the branch or stem. In most cases, the preferred place to make a pruning cut is back to the parent branch or trunk, just to the outside of the **branch collar** or shoulder, if present (Figure 8.1). This position may vary between species, and experience will assist in identification of the correct location. Pruning here most closely simulates where branches are shed naturally. Except on very large branches, the anatomical structure within the branch attachment, called the **branch protection zone**, allows for **compartmentalization** of the wound. This process slows the movement of decay in trees.

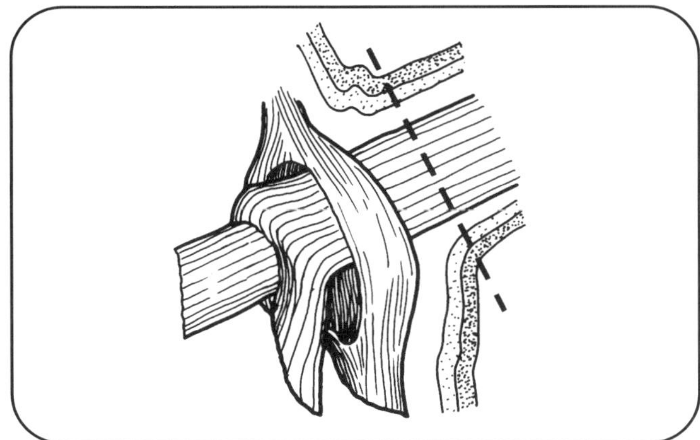

Figure 8.1 Proper pruning cut. Cutting outside the branch collar does not damage trunk tissues.

There are times when it is necessary to reduce the length of a limb by cutting it back to a **lateral** branch. Trees do not respond quite as well to this type of pruning cut because they cannot readily compartmentalize the wound. The ability of the tree to compartmentalize the wound is a function of the size of the cut, the vigor and species of tree, and the climate. The smaller the cut and the more vigorous the tree, the better the closure and compartmentalization. This is important to consider when deciding to use a reduction cut and when deciding how much to remove. When possible, it is best to avoid large cuts of this type. It is also very important to consider the ability of the remaining branch to sustain itself and to assume apical dominance. Cutting back to a lateral that is insufficient in size is much like making a topping cut. Pruning cuts to reduce the length of a branch should bisect the angle between the **branch bark ridge** and an imaginary line perpendicular to the branch or stem being removed (Figure 8.2).

Large or heavy limbs should be removed using three cuts. The first cut undercuts the limb 1 or 2 feet out from the parent branch or trunk. A properly made undercut eliminates the chance of the branch "peeling" or tearing bark as it is removed. The second cut is the top cut, which is usually made slightly farther out on the limb than the undercut. (When cutting large limbs with a chain saw, often the top cut is made directly above the bottom cut to avoid

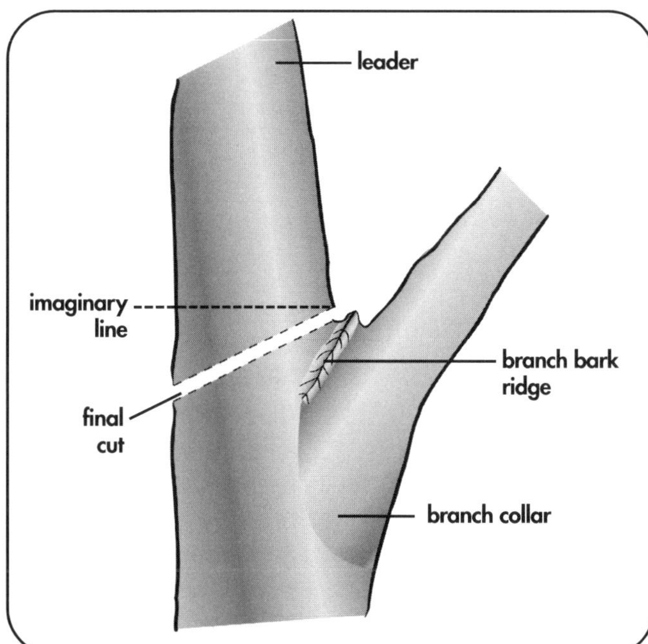

Figure 8.2 When cutting back to a lateral, bisect the angle between the branch bark ridge and an imaginary line perpendicular to the leader or the branch being removed.

Figure 8.4 When removing a dead branch or stub, remove only the dead tissue, and make the final cut just outside the collar of living tissue.

the saw bar getting caught in the kerf.) This allows the limb to drop smoothly when the weight is released. The third cut is to remove the stub (Figure 8.3). When removing a dead branch, the final cut should be made just outside the collar of living tissue. If the collar has grown along a branch stub, only the dead stub should be removed (Figure 8.4).

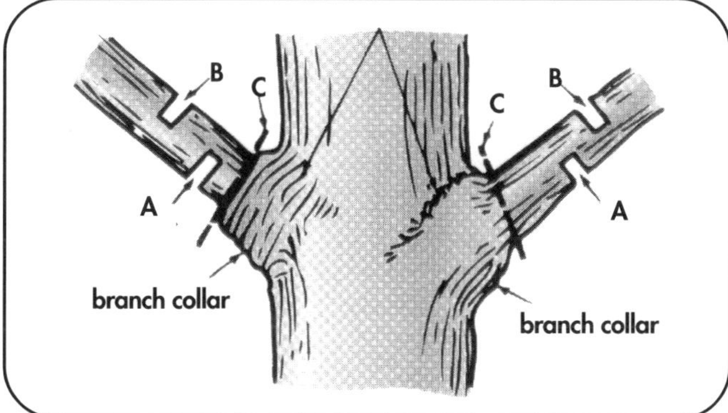

Figure 8.3 Pruning principles. The first cut (A) undercuts the limb. The second cut (B) removes the limb. The final cut (C) should be just outside the branch collar to remove the resultant stub.

STRUCTURAL PRUNING OF YOUNG TREES

Structural pruning principles are used when pruning young trees or a tree that has not been pruned in many years. If young trees are "trained" or pruned to promote good structure, they will likely remain serviceable in the landscape for more years than trees that have not been so pruned. Defects can be removed, a single, dominant leader can be selected, and branches can be well spaced along the main trunk. These trees have a lower potential for structural failure at maturity and require less maintenance later on. Small-maturing, ornamental trees can be trained to several trunks, or pruned to develop only one.

The process of training young trees can be reduced to five simple steps. The first step is to remove broken, dead, dying, or damaged branches. The second step is to select and establish a dominant **leader**. There should be only one leader, which is usually the strongest vertical stem. Competing stems should be **subordinated** (cut back) or removed.

If two branches develop from apical buds at the tip of the same stem, they will form **codominant** stems. Each codominant stem is a direct extension of the trunk (Figures 8.5 and 8.6). It is best if one codominant stem is removed when the tree is young. Branches that have narrow angles of attachment and codominant stems tend to break at the point of attachment, especially if there is **included bark**. Included bark is bark that becomes enclosed inside the crotch as the two branches grow and develop. This weakens the branch attachment, making the tree more prone to failure. Such branches are preferably removed early in the life of the tree. The relative size of a branch in relation to the trunk is more important for strength of branch attachment than is the angle of attachment.

The third step is to select and establish the lowest **permanent branch.** The height of this branch is determined by the location and intended function of

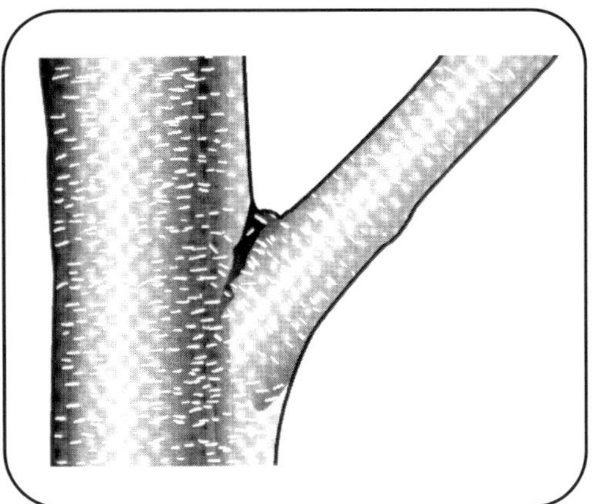

Figure 8.5 Lateral branches should be no more than half the diameter of the parent branch or stem.

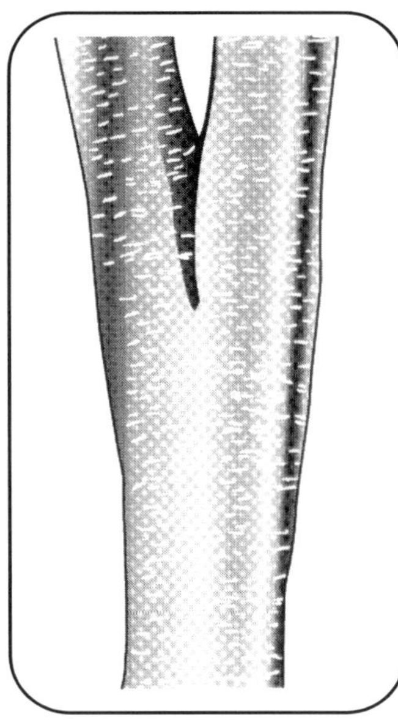

Figure 8.6 Codominant stems are prone to failure as the tree grows larger. One of the stems should be cut back (subordinated) or removed.

the tree. For example, the lowest permanent branch on a street tree should be higher than that on an arboretum specimen. The lowest permanent branch should be less than half the diameter of the trunk at the point of attachment.

Step four is to select and establish **scaffold branches**. These branches should be selected for good attachment, appropriate size, and spacing in relation to other branches. Scaffold branches should be well spaced, both vertically and radially on the trunk. Vertical spacing should be at least 18 inches for large-growing trees, and about 12 inches for smaller trees (Figure 8.7).

The fifth and final step in the process is the selection and subordination of **temporary branches** below the lowest permanent branch and among the scaffold branches. These branches should be retained temporarily because they help provide energy back to the trunk, contribute to trunk taper development, and provide shade to the young trunk tissues. The smaller temporary branches can be left intact; larger ones should be subordinated.

Figure 8.7 Trees should be pruned when they are young to establish a strong, well-spaced branch scaffold, avoiding the poor branch structure shown here.

This training process should be spread out over many years, if practical. Although young trees are often tolerant of severe pruning, the goal should be to remove no more than 25 percent of the canopy in any one year. In most cases, proper training can be accomplished removing much less than 25 percent each year.

PRUNING MATURE TREES

A number of factors must be considered when pruning mature trees. These include the site; time of year; and the species, size, growth habit, vitality, and maturity of the tree. The amount of live tissue that should be removed depends on the tree size, species, and age, as well as the pruning objectives. Younger trees tolerate the removal of a higher percentage of living tissue than mature trees. As a general rule, mature trees are less tolerant of severe pruning than juvenile trees. Also, smaller cuts close faster and are more easily compartmentalized than large cuts (Figure 8.8).

Large, mature trees should require little routine pruning. A widely accepted rule of thumb is never to remove more than one-fourth of a tree's leaf-bearing canopy. In a mature tree, pruning even that much could have negative effects. Removing even a single, large-diameter limb can create a wound that the tree may not be able to close. The older and larger a tree becomes, the less energy it has in reserve to close wounds and defend against decay or insect attack. Further, the

Figure 8.8 Pruning when trees are young allows the arborist to remove smaller branches, leaving smaller wounds. Making large cuts above one another, as shown here, can inhibit the compartmentalization of decay.

energy-producing capacity in relation to mass decreases as a tree matures. The pruning of large, mature trees is usually limited to the removal of dead branches or to reduce the severity of structural defects.

PRUNING TECHNIQUES

Arborists use several different techniques of pruning, depending on the objectives. The most common include crown cleaning, thinning, reduction, restoration, and raising. A few special pruning techniques are used in specific circumstances. The method employed depends on the condition and size of the tree and the wishes of the client. When writing pruning specifications or work orders, a minimum and/or maximum diameter of branches to be removed should be specified with each pruning technique.

Crown cleaning is the selective removal of dead, diseased, broken, or weakly attached branches from a tree crown. This is the most common pruning technique for landscape trees. Regular pruning should correct small growth problems before they have a chance to become large problems.

Crown **thinning** includes crown cleaning as well as selective removal of branches to increase light penetration and air movement through the crown, and to reduce weight (Figure 8.9). Increased light and air stimulates and maintains interior foliage. Thinning toward the tips of a branch can reduce the wind-sail effect of foliar clumps in the crown, and relieve the weight of heavy limbs. Proper thinning should maintain the structural beauty and retain the tree's natural shape. Care must be taken not to over-thin a tree. Clearing out too much inner foliage can have adverse effects on the tree and should be avoided. Vigorous production of **watersprouts** on interior limbs is often a sign of over-thinning.

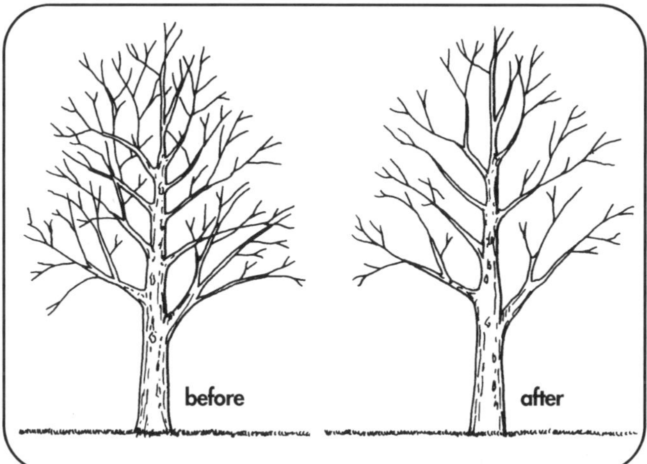

Figure 8.9 Thinning a tree removes unwanted branches, reduces weight, and allows air and light penetration.

When thinning laterals from a limb, an effort should be made to maintain well-spaced inner lateral branches to achieve even distribution of foliage along the branch. Caution must be taken not to create an effect known as **lion tailing**, which is caused by removing an excessive number of inner laterals and foliage (Figure 8.10). This displaces foliar weight to the ends

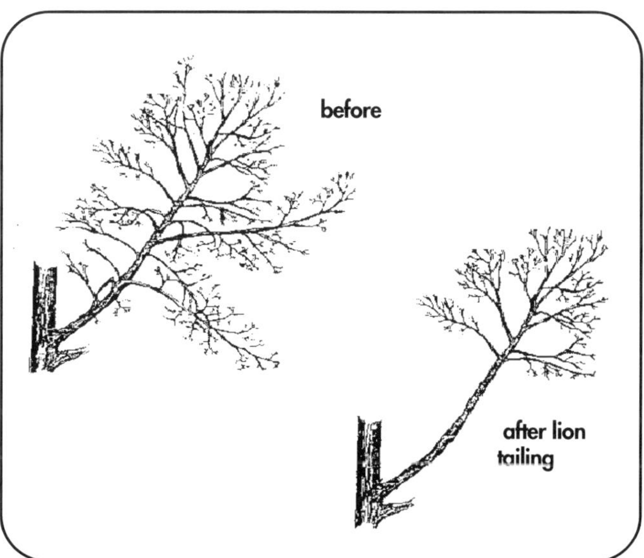

Figure 8.10 Lion tailing makes the limb more prone to breakage. It can be an energy drain on the tree.

of the branches and may result in sunburned bark tissue, watersprouts, reduced branch taper, weakened branch structure, and breakage.

Trees in urban and landscape settings may need to have lower limbs removed. Crown **raising** removes the lower branches of a tree in order to provide clearance for buildings, signs, vehicles, pedestrians, and vistas (Figure 8.11). Excessive removal of lower limbs should be avoided so that development of trunk taper is not affected and structural stability is maintained. Similarly, **vista pruning** is the selective removal or reduction of scaffold limbs to allow a specific view from a predetermined point.

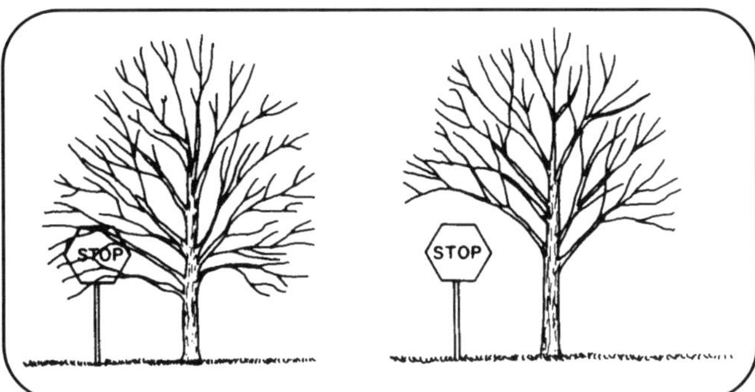

Figure 8.11 Crown raising.

Sometimes the crown of a tree must be reduced in height or spread, such as for utility line clearance. Crown **reduction** is used to reduce the size of a tree. This is best accomplished by cutting limbs back to their point of origin or back to laterals capable of sustaining the remaining limb and assuming apical dominance (Figure 8.12). When a branch is cut back to a lateral, no more than one-fourth of its foliage should be removed. A common rule of thumb is that the remaining lateral branch must be at least one-third the diameter of the removed portion, but this rule varies with species, age, climate, and the condition of the tree. This method will help to maintain the structural integrity and natural form of the tree and delay the time when it will need to be pruned again. Consideration must also be given to the ability of the species to sustain this type of pruning.

Topping or **heading back** involves cutting limbs back to a stub, bud, or a lateral branch not large enough to assume apical dominance. Severe heading causes branch dieback, decay, and sprout production from the cut ends, resulting in a potentially hazardous situation once the sprouts become large and heavy. Topping or heading back is not a recommended pruning practice (Table 8.1 and Figures 8.13 and 8.14).

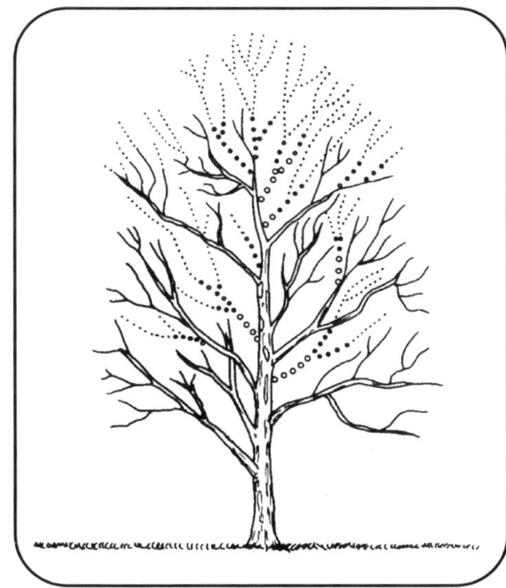

Figure 8.12 If the height of a tree must be reduced, all cuts should be made to strong laterals or to the parent limb. Do not cut limbs back to stubs or to laterals not large enough to assume apical dominance.

If a tree has been topped previously or has sustained storm damage and has sprouted vigorously, crown **restoration** can improve its structure and appearance. Restoration consists of the selective removal of some watersprouts, all stubs, and dead branches to improve a tree's structure and form. One to three sprouts on main branch stubs are selected to become permanent branches and to reform a more natural-appearing crown. Selected vigorous sprouts may need to be subordinated to control length growth and ensure adequate attachment for the size of the sprout. Restoration usually requires several prunings over a number of years.

Utility pruning is the removal of branches or stems to prevent the loss of service, prevent damage to utility

Figure 8.13 Topping causes sprout production below the cuts. These sprouts are weakly attached.

Table 8.1 Why *Not* to "Top"—Eight Good Reasons

1. **Starvation:** Good pruning practices rarely remove more than one-fourth of the crown, which in turn does not seriously interfere with the ability of a tree's leafy crown to manufacture food. Topping removes so much of the crown that it upsets an older tree's well-developed crown-to-root ratio and temporarily cuts off its food-making ability.
2. **Shock:** A tree's crown is like an umbrella that shields much of the tree from the direct rays of the sun. By suddenly removing this protection, the remaining bark tissue is so exposed that scalding may result. There may also be a dramatic effect on neighboring trees and shrubs. If the tree thrives in shade and the shade is removed, poor health or death may result.
3. **Insects and Disease:** The large stubs of a topped tree have a difficult time forming callus. The terminal location of these cuts, as well as their large diameter, prevent the tree's chemically based natural defense system from doing its job. The stubs are highly vulnerable to insect invasion and the spores of decay fungi. If decay is already present in the limb, opening the limb will speed the spread of the disease.
4. **Weak Limbs:** At best, the wood of a new limb that sprouts after a larger limb is truncated is more weakly attached than a limb that develops more normally. If rot exists or develops at the severed end of the limb, the weight of the sprout makes a bad situation even worse.
5. **Rapid New Growth:** The goal of topping is usually to control the height and spread of a tree. Actually, it has the opposite effect. The resulting sprouts (often called watersprouts) are far more numerous than normal new growth, and they elongate so rapidly that the tree returns to its original height in a very short time—and with a far denser crown.
6. **Tree Death:** Some older trees are more tolerant of topping than others. Beeches, for example, do not sprout readily after severe pruning, and the reduced foliage most surely will lead to death of the tree.
7. **Ugliness:** A topped tree is a disfigured tree. Even with its regrowth, it never regains the grace and character of its species. The landscape and the community are robbed of a valuable asset.
8. **Cost:** To a worker with a saw, topping a tree is much easier than applying the skill and judgment of good pruning. Therefore, topping may cost less in the short run. However, the true costs of topping are hidden. These costs include reduced property value, the expense of removal and replacement if the tree dies, the loss of other trees and shrubs if they succumb to changed light conditions, the risk of liability from weakened branches, and increased future maintenance.

(Courtesy of The National Arbor Day Foundation)

Figure 8.14 Stubs left from topping are likely to decay.

equipment, avoid impairment, and uphold the intended usage of utility facilities. Only qualified line-clearance tree trimmers or qualified line-clearance trainees should engage in line-clearance work. It is sometimes necessary to prune trees outside the scope of landscape pruning guidelines to accomplish these objectives. When practical, however, cuts should be made in accordance to A300 standards, and the natural shape and structure of the tree should be maintained, if practical.

PRUNING PALMS

Palm pruning should be performed when **fronds**, flowers, fruit, or loose petioles may create a dangerous condition. It is preferable not to remove live, healthy fronds. If they must be removed, however, avoid removing fronds that initiate at an angle of 45 degrees or greater above horizontal. Fronds removed should be severed close to the petiole base without damaging living trunk tissue.

SPECIALTY PRUNING

Sometimes pruning is conducted to create special aesthetic effects. **Espalier** is a combination of cutting and training branches that are oriented in one plane, usually supported on a wall, fence, or trellis. The pattern can be simple or complex, formal or informal. This technique has been used for centuries on fruit trees, often when space was extremely limited.

Pollarding is a training system that involves severe heading the first year, and sprout removal annually or every few years to keep large-growing trees to a modest size or to maintain a formal appearance. Because of the abundance of sprouts that a tree produces in this process, it is thought that the technique was formerly used for shoot generation for fuel, shelter, and various products. The pollarding process should be started when the tree is young. It is not an effective technique on every species of tree (Figure 8.15).

Internodal cuts are made at specific locations to begin the pollarding process. After the initial cuts are made, no additional internodal cuts should be made. Pollard heads, called knobs, will develop at these points, and the tree will produce sprouts from these knobs. Sprouts that grow from these knobs should then be removed annually or every few years, during the dormant season, taking care not to cut into or below the knobs. The knobs are the key differentiating factor between pollarding and topping. If damaged or removed in subsequent pruning, the branches will react as they would on a topped tree.

WOUND DRESSINGS

Wound dressings were once thought to accelerate wound closure and reduce decay. Research has not substantiated this, however. Some studies have shown beneficial effects in specific cases in reducing borer attack, oak wilt infection, or control of sprout production or mistletoe. Wound dressings are used primarily for cosmetic purposes, and are neither required nor recommended in most cases. If a dressing must be applied, only a light coating of a non-phytotoxic material should be used.

Figure 8.15 Pollarding is a labor-intensive technique that is not appropriate for all species or all sites.

PRUNING TOOLS

Pruning tools adequate for the size of cuts being made should be selected. Tools should be sharp so as to make clean cuts without jagged edges or stubs. Anvil-type pruning tools, with a blade that cuts to a flat surface, should be avoided; tools with bypass blades are preferred. Equipment and work practices that damage living tissue and bark beyond the scope of the work should be avoided. Climbing spurs should not be used to climb trees for pruning operations.

Some instances call for sterilization of pruning tools between plants, or even between cuts. The goal is to minimize the chance of spreading certain diseases. The probability of spreading diseases in this fashion varies with the particular disease, the plant, the pruning tools used, the environmental conditions, and the timing. If tools are sterilized, it is important to use a material that will not damage plant tissues.

PLANT GROWTH REGULATORS

Plant growth regulators are substances, usually effective in small quantities, that enhance or alter the growth and development processes of a plant. In most cases, these chemicals either increase or decrease normal growth, flowering, or fruiting of plants.

Utility arborists sometimes use growth regulators to control the growth of trees and other vegetation near or beneath utility lines. Growth inhibitors can be sprayed on the foliage, banded on the bark, soil applied, or injected into the tree. Some of the growth regulators (**antigibberellins**) inhibit the synthesis of the naturally occurring cell-elongation hormone, gibberellin. These chemicals can significantly reduce pruning expenses. There is a great deal of research being done to find ideal formulations, while minimizing phytotoxicity.

Another use of growth regulators is to reduce watersprout production on trees. Studies have shown that watersprout and sucker growth can be minimized, in certain instances, with the use of growth regulators. Interest in the use of plant growth regulators is increasing among utility arborists, though they have not gained widespread acceptance by commercial arborists. Concern over the use of chemicals in the landscape may be a limiting factor.

Chapter 8 Workbook

1. Because pruning removes leaves and reduces the overall photosynthetic capacity of a tree, it also reduces __________.
2. Name five common reasons for removing limbs from trees.
   a.
   b.
   c.
   d.
   e.
3. To maximize flowering, plants that bloom on the current season's wood should be pruned prior to ______ ______________________, or in the summer after bloom has occurred. Plants that bloom on last season's wood should be pruned just after __________.
4. True/False—Trees that tend to "bleed" should never be pruned in the early spring because this is likely to cause a major decline in vigor.
5. Label the branch bark ridge and the branch collar on this drawing. Show where the undercut, top cut, and final cut should be made in removing the limb.

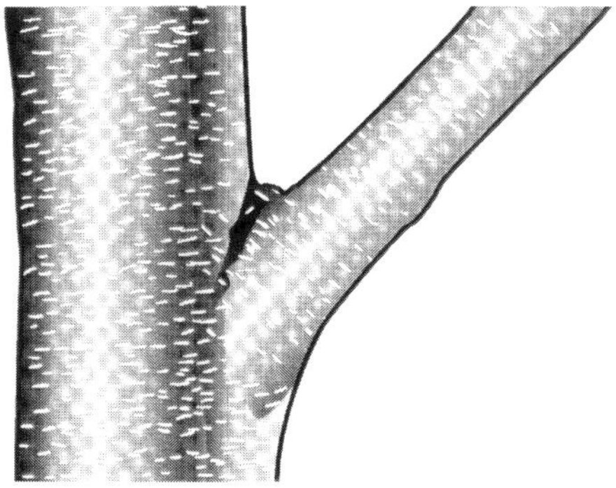

6. ________________ __________ is bark that gets pushed inside a crotch as two branches grow and develop.
7. The swollen area at the base of a branch where it arises from the trunk is called the ______________ _____________.
8. Two limbs that arise from apical buds on the same stem are known as _________________ ______, and are both direct extensions of the stem below.
9. True/False—In the absence of included bark, the relative size of a branch in relation to the trunk is more important for strength of branch attachment than is the angle of attachment.
10. True/False—Codominant stems can represent a structurally unstable branch configuration, especially if there is included bark in the junction.
11. When training young trees, a single, central leader should be selected, and competing leaders should be removed or ____________________.
12. The presence of ________________ __________ in a crotch weakens branch attachment.

13. When practical, temporary lower branches should be left on a young tree to help develop trunk ____________.
14. Avoid removing more than ______ percent of the canopy in a given year.
15. ____________ ______________ is the removal of dead, dying, and weak branches from a tree.
16. True/False—Removing even 25 percent of the canopy of a large, mature tree can be stressful to the tree.
17. _________________ includes crown cleaning as well as selective removal of branches to increase light penetration and air movement into the crown of the tree.
18. Caution must be taken not to create an effect known as ______ ______________, which is caused by excessive removal of inner laterals and foliage.
19. Three adverse effects of lion tailing are
    a.
    b.
    c.
20. _________________ is best accomplished by cutting limbs back to laterals that are large enough to sustain the remaining branch and assume the terminal role.
21. Three adverse effects of topping or heading back include
    a.
    b.
    c.
22. True/False—In general, removal of more than one-third of a tree's crown in a single pruning is acceptable as long as the tree is vigorous.
23. True/False—Research on wound dressings shows that their use prevents decay.
24. _______ ___________ ____________ are substances, usually effective in small quantities, that enhance or alter the growth and development of a plant.
25. Name two ways that plant growth regulators are used by arborists.
    a.
    b.

CHALLENGE QUESTIONS

1. What are the effects of pruning on the growth and development of roots and shoots of a tree?

2. Explain the response of a tree to topping. What steps should be taken to restore the crown? What alternatives might be suggested to a client who wants his or her tree topped?

3. Why is good trunk taper important in the development of a tree?

4. How do plant growth regulators reduce the growth of trees? What are the limitations to their use?

SAMPLE TEST QUESTIONS

1. When pruning young trees, it is important to train for a dominant leader and well-spaced scaffold branches so that
   a. future pruning can be minimized
   b. the tree will be structurally strong
   c. codominant branching can be avoided
   d. all of the above
2. To prune trees that flower on the previous year's growth and to maximize flowering, you should
   a. prune any time during the dormant season
   b. prune shortly after flowering
   c. prune in late summer after seed formation
   d. prune in the fall, just after leaf drop
3. When pruning a branch from a tree, the final cut should be
   a. flush with the parent stem
   b. at a 45-degree angle to the parent stem
   c. parallel to the branch bark ridge
   d. just outside the branch collar
4. When it comes to pruning, as a rule, mature trees are
   a. more tolerant of extremes than young trees
   b. capable of tolerating heading
   c. not tolerant of severe pruning
   d. unlikely to produce watersprouts
5. If the height of a tree must be reduced,
   a. branches should be removed at their point of origin or to a lateral large enough to assume apical dominance
   b. all cuts should be made at internodes to avoid cutting through buds
   c. the tree should be root pruned to compensate for foliage loss
   d. all of the above

Other Sources of Information

(See pages v–vi for complete bibliographic information.)

ANSI A300. *Standard Practices for Tree, Shrub, and Other Woody Plant Maintenance, Part 1: Pruning.*
Costello, 2000. *Training Young Trees for Structure and Form.*
Gilman, 1997. *An Illustrated Guide to Pruning Trees.*
ISA, 1995. *Tree-Pruning Guidelines.*

CHAPTER 9

TREE SUPPORT AND PROTECTION SYSTEMS

CHAPTER 9 TREE SUPPORT AND PROTECTION SYSTEMS

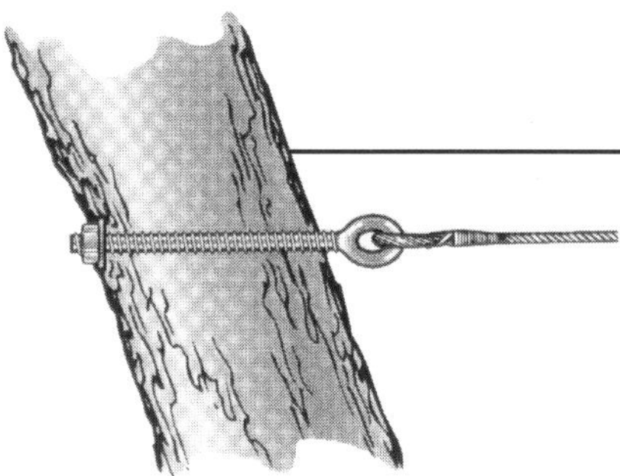

CHAPTER 9
TREE SUPPORT AND PROTECTION SYSTEMS

Objectives

1. Understand when a tree might be helped by the installation of cables, guys, or bracing rods.
2. Recognize the limitations of cabling and bracing systems.
3. Explain the different ways trees can be cabled, braced, or guyed.
4. Become familiar with the techniques and materials used in cabling and bracing.
5. Recognize that by cabling, bracing, or guying a tree, the arborist takes on a future responsibility to a client. Periodic inspection of cables is required to ensure proper functioning of the cable system.
6. Understand the circumstances under which a lightning protection system may be recommended for trees.
7. Learn the techniques and materials used in the installation of lightning protection for trees.

Key Terms

7-strand, common-grade cable
air terminal
amon-eye nuts
box cable system
cable aid
cable clamps
cable grip
come-along
dead-end grips
dead-end hardware
direct cable system
extra-high-strength cable
eye bolts
eye splice
ground rod
guying
lag eye
lag hook/J-hook
lag-threaded rod
machine-threaded rod
main conductor
ship auger
thimble
threaded rods
triangular cable system

INTRODUCTION

Tree cabling, bracing, guying, and lightning protection each involves the installation of hardware in trees. Any time hardware is installed in a tree, there will be wounding and the risk of decay.

Hardware such as cables or braces is installed in trees to provide extra support by limiting the movement of limbs. When used wisely, they may reduce the risk of failure and extend the life of the tree. In determining whether cabling is warranted, the condition of the tree should be considered. If the root system is not structurally sound, or if the tree contains excessive decay, removal of the tree may be preferable. Hazardous trees cannot be made safe by the use of cables. Therefore, the situation must be assessed carefully when deciding to install supplemental support systems, to prune potentially hazardous or decayed limbs, or to remove the tree.

Lightning protection involves the installation of hardware for a very different purpose. It is intended to direct the electrical charge of a lightning strike safely away from the tree. Again, the tree should be assessed before determining whether it is a good candidate. If a system is installed, provisions should be made for future inspections and maintenance.

CABLE INSTALLATION

Cables are installed in trees to limit movement of limbs judged to have a weak connection or to support heavy limbs by connecting two or more limbs together. Additional support may be needed due to split or decayed crotches, crotches with included bark, or the inherent danger of weak-wooded trees (Figure 9.1). Multi-stemmed trees are susceptible to breaking under the stress of wind or the weight of accumulated ice or snow. Branches that pose a potential threat to property or people may be candidates for cabling. In determining whether cabling is warranted, the condition of the tree should be assessed. If the root system is not structurally sound, or if the tree contains excessive decay, removal of the tree may be preferable. Support cables, even when combined with braces, have limitations. Mechanical devices cannot always be relied upon to make a potentially hazardous tree safe.

Before installing cables or braces in a tree, the tree should be properly pruned. If necessary, the tree should be pruned to remove hazardous limbs, and to reduce the weight of the limbs to be cabled.

Figure 9.1 Splits such as this may be candidates for support with a cable and a brace rod.

Cabling Hardware

Cables

It is important to select the appropriate hardware for cabling a tree. Cables, eye bolts, and other cabling hardware come in various sizes and types. Consider the size of the limbs, the weight to be supported, and the presence of decay when choosing materials. If the hardware is too small or inadequate, the cable may fail. The ANSI A300 standards for tree support systems specify minimum hardware sizes for various sizes of limbs.

Two types of steel cable are commonly available for use in cabling trees: **7-strand, common-grade galvanized** and **extra-high-strength (EHS)** cable. The common-grade cable is relatively malleable (bendable) and easy to work with. The EHS cable is much stronger but less flexible than common-grade cable. Both are available in a wide range of sizes. Cables 3⁄16-inch diameter to 3⁄8-inch diameter are commonly used in trees.

Dead-End Hardware

A **lag eye** is a lag-threaded, drop-forged anchor with a closed eye. A **lag hook**, or J-hook, is a lag-threaded, drop-forged device in the shape of a "J." Lag hooks are J-shaped, with the long end threaded with wood screw threads (Figure 9.2). Standard lag sizes used in tree cabling are 5⁄16-inch, 3⁄8-inch, 1⁄2-inch, and 5⁄8-inch diameter. Lag hooks come with right- and left-handed threads, so that when each end is twisted into the branch to tension the cable, the cable will not come unlaid or unwound.

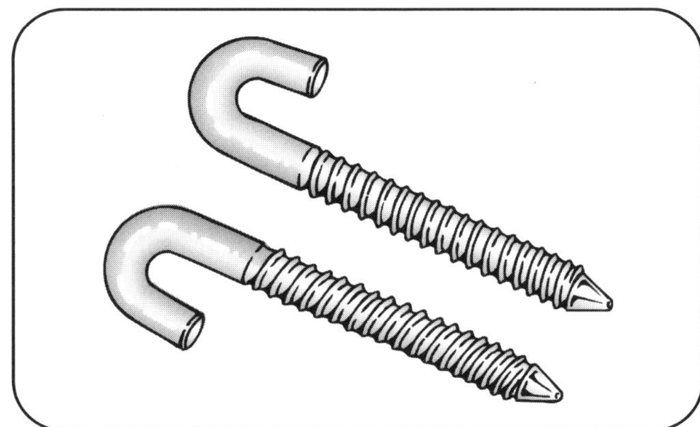

Figure 9.2 Left and right lag hooks.

Lag hooks are installed by screwing into a predrilled hole that is smaller in diameter than the lag. The rule of thumb is to drill the hole 1⁄16- to 1⁄8-inch smaller than the lag. The lags should be screwed into the tree so that the J-loop ends up vertical (facing up or down), with the open end just contacting the bark. Care should be taken to avoid injuring the bark. If the lag must be installed at an angle that will prevent screwing it in completely, it may be preferable to install an eye bolt or other anchoring hardware.

Dead-end hardware works quite well on small limbs and trees with hard wood, but should not be installed in limbs that are greater than 8 to 10 inches in diameter. Dead-end hardware should never be installed in limbs with decay. The decay will limit the holding capacity of the lags and may spread the decay into healthy wood. Lags should not be used if it will not be possible to seat the full length of the threads.

Through-Hardware

In circumstances where dead-end hardware is not appropriate, **eye bolts** or **threaded rods** with **amon-eye nuts** may be used (Figure 9.3). Both are drop forged and machine threaded, and their installation is similar. A hole 1⁄16- to 1⁄8-inch larger than the hardware is drilled through the limb to be cabled. The eye bolt or threaded rod is installed with a round washer and nut on the outside end. The washer should be seated against the bark. On trees with very thick bark, the bark should be chiseled away or drilled to countersink the washer against the sapwood. The exposed threads on the end of the eye bolt, and both ends of the threaded rod used with the amon-eye, should be peened to prevent the nut from unscrewing.

Drop-forged eye bolts are considered slightly stronger than threaded rods used with amon-eye nuts. However, an advantage of the amon-eye system is that the length of the rod can be adjusted easily for any job.

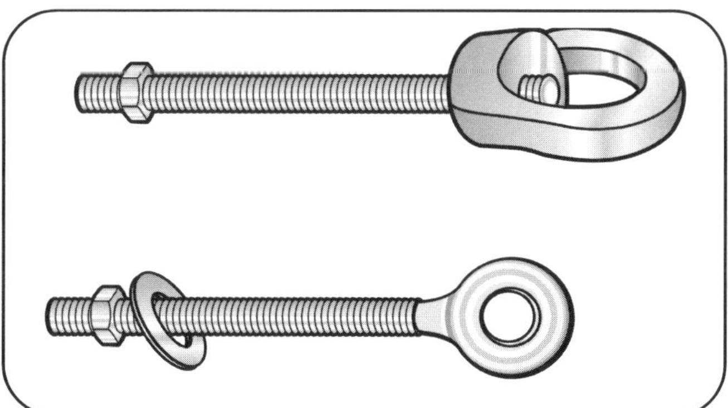

Figure 9.3 Threaded rod with amon-eye nut (top); eye bolt (bottom).

Washers, Thimbles, and Dead-End Grips

If eye bolts or threaded rods are used to install cables, they must be anchored with washers and nuts. Heavy-duty, round washers are recommended, although oval washers are acceptable. Heavy-duty washers are larger in diameter and thicker than standard washers. The diamond-shaped washers that were formerly recommended are no longer considered acceptable. They tend to slow closure over the wound and may cause vertical cracks.

When attaching the cable to its anchoring hardware, **thimbles** must be used. The purpose of the thimble is to protect the cable from excessive wear and increase the bend radius. If soft, common-grade cable is installed directly on the hardware, the steel-to-steel contact and abrasion may eventually cause wear and cable breakage.

Thimbles must also be used when installing **dead-end grips** (Figure 9.4). Dead-end grips are installed over a heavy-duty thimble and then wrapped on the cable. They are used to attach EHS cable to hardware. The cable is not malleable enough to form an eye splice.

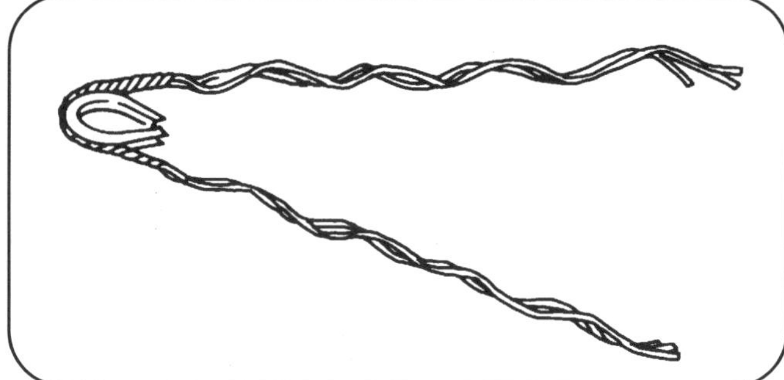

Figure 9.4 Dead-end grip.

Dead-end grips have been found to be quite satisfactory for small- and medium-sized trees. However, they are not designed to withstand dynamic loads that can be created by branches twisting and swaying, particularly in gusting winds. Excessive wind sway in large trees may cause metal fatigue, which could lead to failure. When installing this hardware in large trees, it is important to take into account the wind conditions and potential for twisting.

Cabling Tools and Equipment

The installation of cables in trees requires a number of tools. Often a **come-along** is used to bring two branches closer together (Figure 9.5). A second commonly used tool is a **cable grip.** This is a device used to grip the cable and help the climber pull the cable for tensioning or attaching to anchor hardware.

Figure 9.5 The limbs may be brought closer together using a come-along to ensure the cable will be taut.

Grips used to tension the cable must be the correct type designed for use with the cable being installed. Another handy piece of equipment is the cable aid. A **cable aid** can be used to spread open thimbles, to tighten lags, and to help wrap dead-end grips onto the cable.

To install the hardware in the tree, holes must be drilled (Figure 9.6). This can be accomplished using either a brace and bit, or a power drill. Though fast and efficient, electric drills can be difficult to work with in the tree, and an electrical source is not always available. Gas-powered or battery-powered, rechargeable drills offer a solution to these problems. Whether using a power drill or a brace and bit, the bit should be a **ship auger**, which will work more efficiently in green wood and will pull the shavings from the hole (Figure 9.7).

Other tools useful in cabling operations include cable cutters, hacksaw, hammer, chisel and mallet, slings, and wire cutters. It is helpful for a climber to carry these tools in a bag, bucket, or belt that will prevent them from being accidentally dropped.

Attaching the Cable to the Hardware

As previously mentioned, if EHS cable is used, it must be attached to the hardware using dead-end grips. However, if common-grade cable is used, there are alternatives. Seven-strand, common-grade cable is often attached to the hardware using an **eye splice**. An eye splice is actually a series of wraps, and not a true splice. A thimble is always used to form the eye in the end of the cable (Figure 9.8). Remember to use the right size of thimble for the cable. If the bend in the thimble is too tight, the cable will be weakened. The eye splice is made by wrapping the end of the cable around the thimble, then separating the cable strands. Each strand is wrapped individually around the cable with at least two complete turns per strand in the same direction. When complete, the finished splice will have a neat appearance and will provide optimal hold (Figure 9.9).

Figure 9.6 Drilling for cable installation using a power drill.

Figure 9.7 Overhead view of a climber drilling for eye-bolt installation using a gasoline-powered drill.

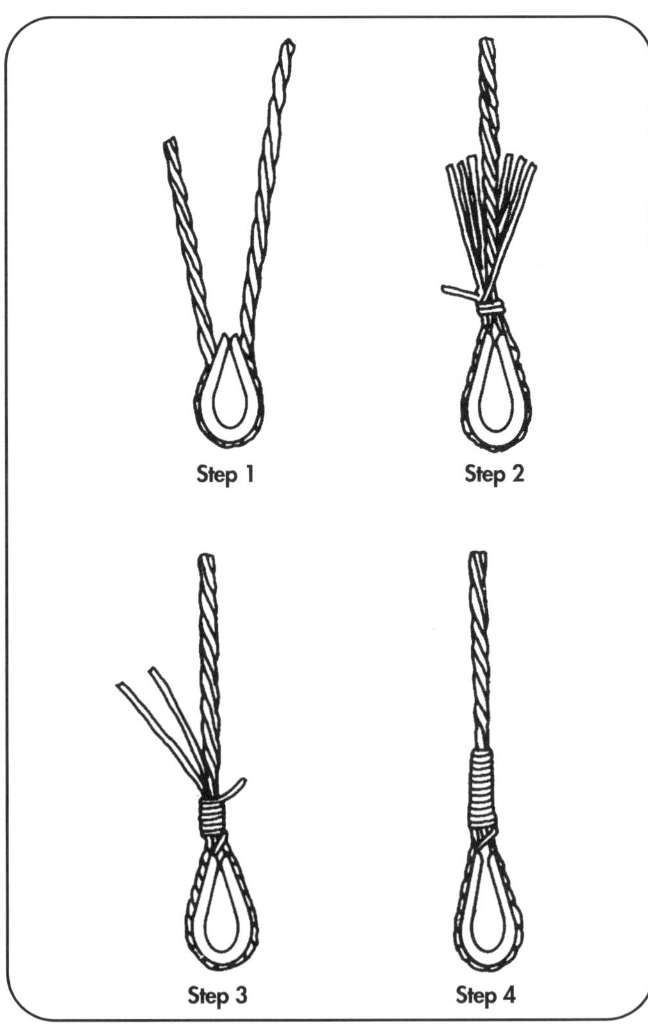

Figure 9.8 Forming an eye splice.

Figure 9.9 Forming the eye splice on the cable.

An alternative to forming an eye splice is to use **cable clamps** (wire rope clamps). These clamps should not be used to form terminations on cables greater than ⅛-inch diameter. It is imperative that cable clamps be installed correctly if they are to offer reliable anchorage. If cable clamps are installed, three should be used for adequate strength. They must be installed several inches apart, with the "U" of the clamp over the short end of the cable. Many arborists prefer not to use cable clamps, believing that other means of termination offer more reliable anchorage and a more professional appearance.

Installation Techniques

A general rule of thumb is to install the cable at least two-thirds the distance from the defect or weak crotch to the ends of the limbs. Exact placement depends on the location of lateral branches and defects. The branches at the point of cable attachment must be large enough and solid enough to provide adequate support of the hardware (Figure 9.10).

The angle of the cable and its distance from the crotch determine its strength and effectiveness. The support can be maximized by installing the cable directly across the crotch being supported, and at least two-thirds up. "Directly across" can be determined by setting the cable perpendicular to (at a 90-degree angle to) an imaginary line through the center of the crotch (Figure 9.11).

The hardware should be installed with the pull of cable in direct line with the installed bolt or lag. If the cable is installed correctly, at right angles with the line that bisects the crotch, the hardware will usually ***not*** be installed perpendicular to either branch. To maximize the strength of the system, it is important to have the cable's pull in direct line with the hardware, not at an angle (Figure 9.12).

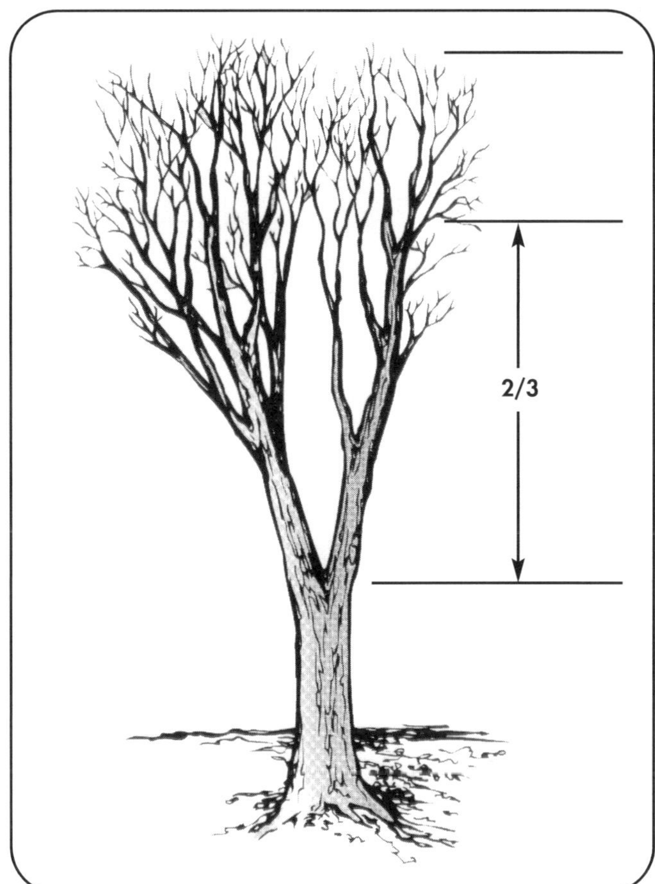

Figure 9.10 Cables should be installed at least two-thirds the distance from the crotch to the branch tips.

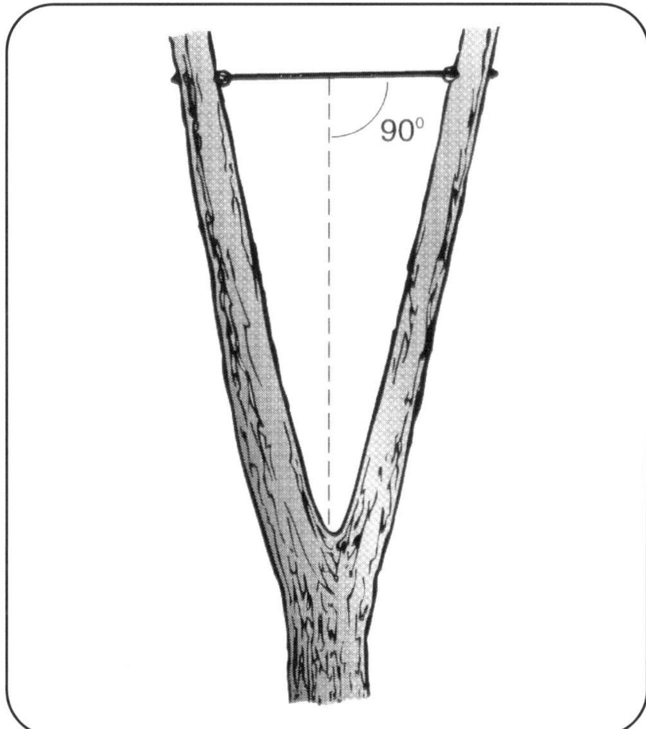

Figure 9.11 The cable should be installed perpendicular to an imaginary line that bisects the crotch.

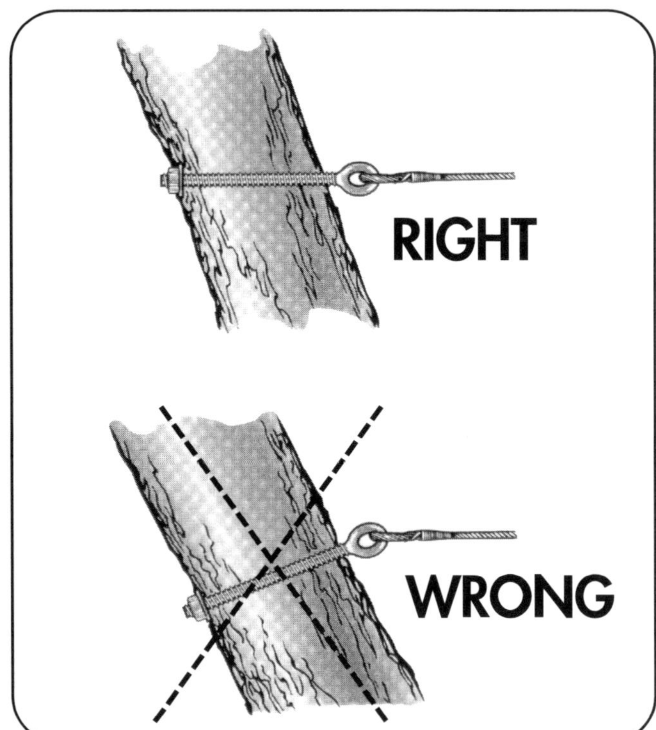

Figure 9.12 The hardware should be installed in direct line with the pull of the cable.

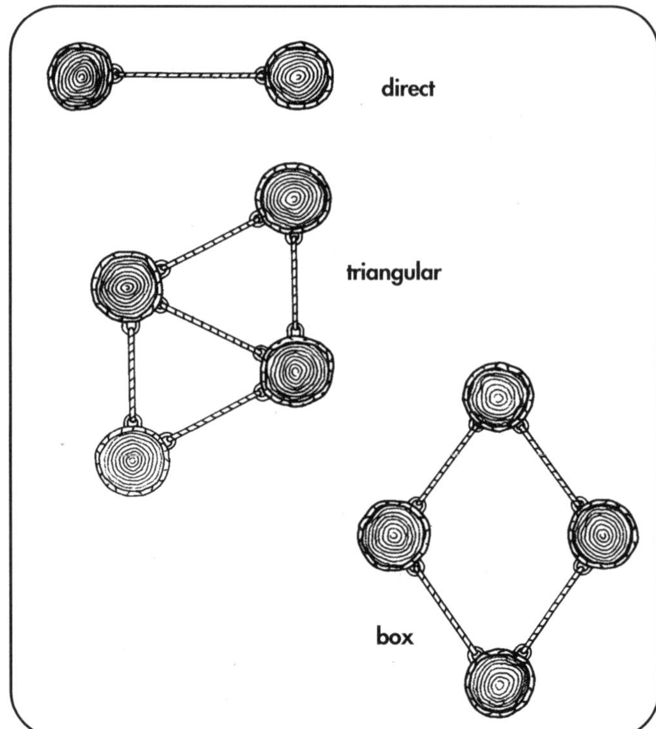

Figure 9.13 Cabling systems.

Once installed, the cable should be just taut. A cable that is too tight may put excessive stress on the wood fibers, result in more damage at the defect, or cause the hardware to pull out. The limbs may be brought closer together with ropes, or slings and a come-along. This will make the installation easier, and the cable should be taut when the limbs are released. If the cable is installed while the tree is in foliage, it should be tight enough that it will not be slack after the leaves have fallen.

The most common cable installation is the simple or **direct cable** (one cable between two limbs). Sometimes a tree will require more than one cable. Several cabling systems are illustrated in Figure 9.13. If multiple cables are required, extra strength can be added to the system by cabling the branches together in threes (triangular). If more crown movement is desired, a box or rotary system can be installed.

When more than one cable is being installed on the same limb, hardware should be spaced no closer together than a distance equal to the diameter of the limb, and never in longitudinal alignment with other hardware (Figure 9.14). Internal decay from the hardware installation could coalesce, further weakening the support. Cables must not rub against wood or each other. Only one cable should be attached to any one anchor (Figure 9.15).

Anchors should not be installed into decayed areas where the amount of sound wood is less than 30 percent of the trunk or branch diameter. The percentage of sound wood can be estimated using the equation in Figure 9.16 if the diameter of the decayed portion is known. If the decayed portion is too large, removal of the limb or tree should be considered. Remember that most of the holding power of the bolt or lag is from the wood and callus that forms along the hardware after installation (Figure 9.17). The wood inside the branch at the time of installation may decay further, thus reducing holding capacity.

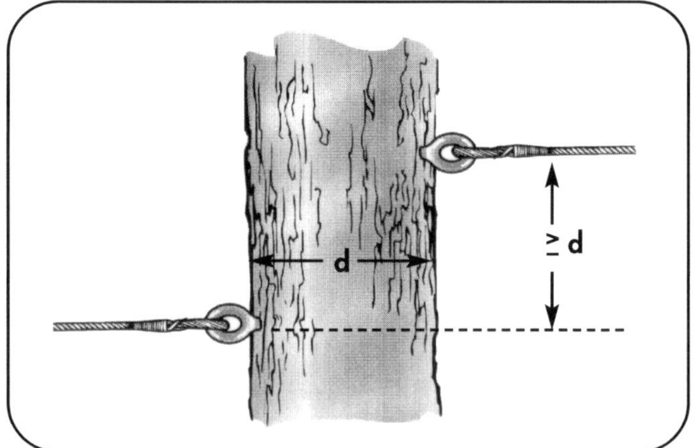

Figure 9.14 Hardware installed in trees should be no closer together than a distance equal to the diameter of the limb.

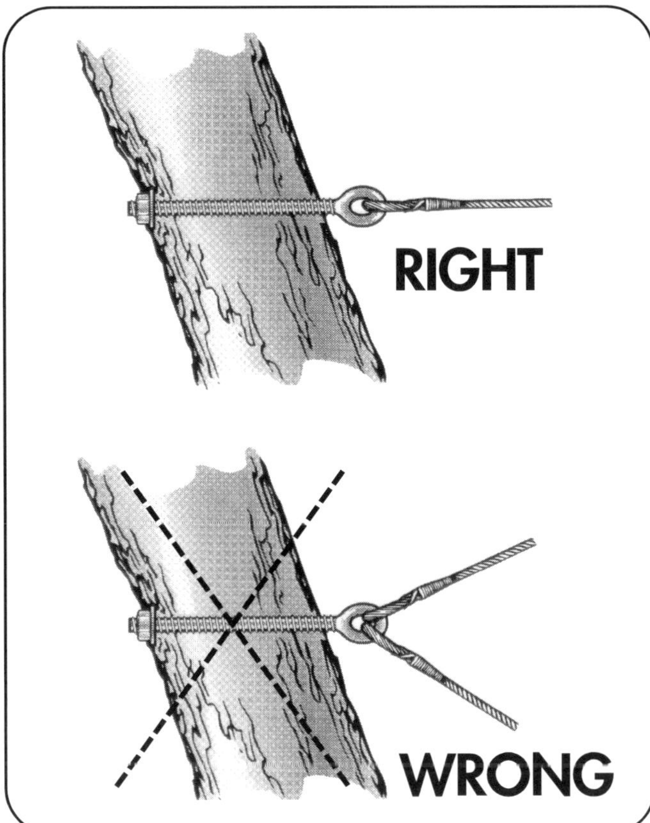

Figure 9.15 Only one cable should be attached to each anchor.

The installation of cables in a tree represents an ongoing responsibility. Cables must be inspected annually. Periodic inspections should be made to check the structural integrity of the tree and limbs, the condition of the hardware and support system, the cable tension, and the position of the cable(s) in the tree. As the tree grows older and taller, new cables may eventually need to be installed higher in the tree. Old systems should not be removed until after the proper installation of the new system. Trees that have been cabled may also need to be pruned periodically to remove excess foliage weight and reduce wind resistance.

Nonrigid Support Systems

For decades, arborists have been installing steel cables in trees to provide added support. More recently, nonrigid support systems have been developed. One common system uses polypropylene rope that is wrapped around the limbs over a wide band to distribute forces, and through a protective sheath to reduce friction on the rope. The ends of the rope feed back into the core, not unlike a Chinese finger-trap device. A rubber insert is also fed into the core of the rope to provide a shock-absorbing effect. The idea is to reduce the shock loading created when the cabled limbs sway back and forth in the wind. One potential advantage is that these nonrigid systems do not require drilling to anchor hardware. More research is required, however, to determine the effects of UV-light degradation of the polypropylene, the constriction effects, if any, on the limbs, and the strength of the system years after installation.

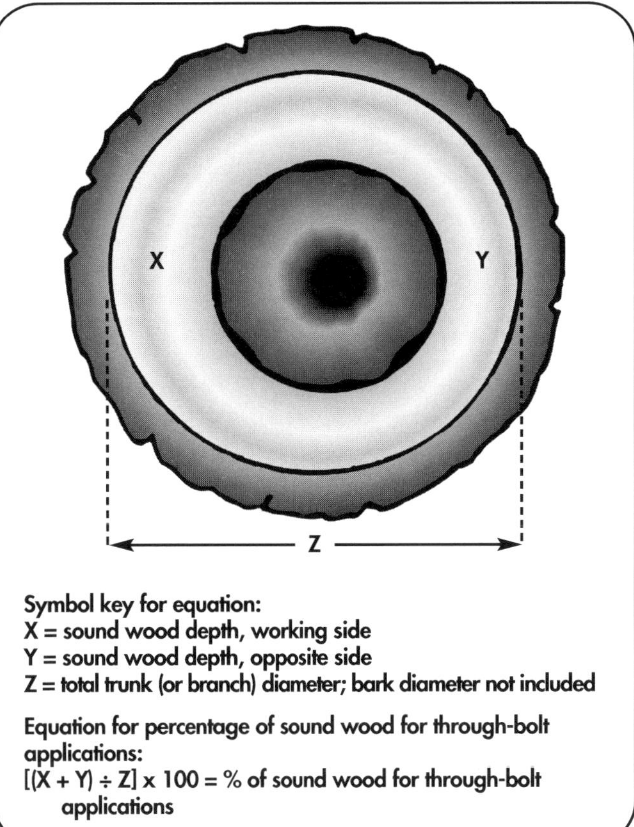

Symbol key for equation:
X = sound wood depth, working side
Y = sound wood depth, opposite side
Z = total trunk (or branch) diameter; bark diameter not included

Equation for percentage of sound wood for through-bolt applications:
[(X + Y) ÷ Z] x 100 = % of sound wood for through-bolt applications

Figure 9.16 Equation for finding the percentage of sound wood.

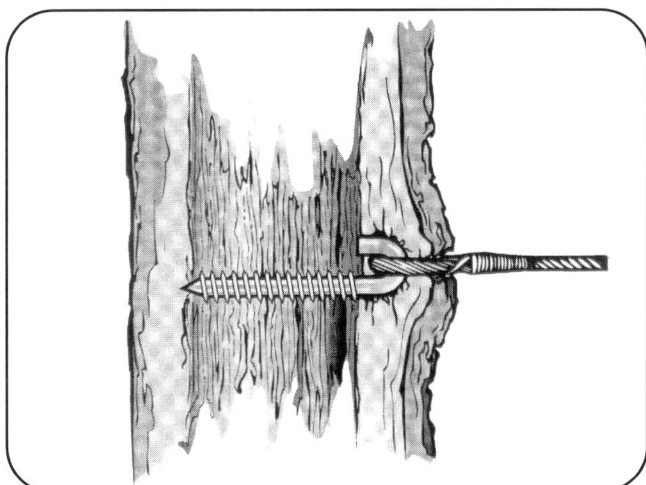

Figure 9.17 Much of the holding power comes from wood that forms after installation. At this point, branch twisting can cause metal fatigue, however.

BRACING

Bracing is the use of steel rods in limbs, leaders, or trunks to provide rigid support for a tree. Bracing is used to reinforce weak or split crotches, or to strengthen decayed areas. In most cases, bracing is used in combination with cabling, not as a substitute.

There are two types of steel rods used in tree bracing: **lag-threaded rods** and **machine-threaded rods**. Lag-threaded rods have fewer and deeper threads per inch than machine-threaded rods. Lag-threaded rods can be installed as dead-end hardware, with the hole drilled 1/16- to 1/8-inch smaller than the diameter of the rod. If the rod is installed through two sections of a tree, the hole is drilled completely through the first (smaller) section, and at least halfway through the second. The rod is then threaded into the tree. The rod is scored for breaking at a distance that will allow the break to be inside the cambium of the tree. The rod is threaded the rest of the way in, and the excess rod is broken off. The advantage to using lag-threaded rods is that the threaded rod itself provides support. Several rods may be used in combination without outside hardware.

In large trees, soft-wooded trees, or decayed wood, a machine-threaded rod should be used. In this case, the hole is drilled 1/16- to 1/8-inch larger than the diameter of the rod. It is necessary to use a drill bit of sufficient length to penetrate all the way through the section(s) being braced. Never attempt to drill from opposite ends and hope to have the holes align. The rod is fed through the tree and bolted on each end using round or oval, heavy-duty, heat-treated washers and nuts (Figure 9.18). The washers should be seated on the bark. Excess rod beyond the nuts should be cut off, and the end of the rod should be peened to prevent the nuts from backing off (Figure 9.19). Any portion of the steel rod that is exposed should be treated with rust-preventive paint.

If a single rod is being installed to support a crotch that is not split, it should be installed above the crotch, at a distance of about one to two times the branch diameter above the crotch. Additional braces may be installed below the crotch, if necessary. Bracing for split crotches should be installed below the crotch and through the split.

In many cases, it may be advisable to install more than one bracing rod. Multiple rods provide added strength and reinforcement to a split or weak crotch. If more than one bracing rod is to be installed, they should be staggered. Longitudinal alignment of the rods may add to the formation of decay columns in the tree. Whenever practical, the rods should be spaced apart a distance greater than the diameter of wood being braced (Figure 9.20). To provide added support, cables should also be installed in trees that have been braced.

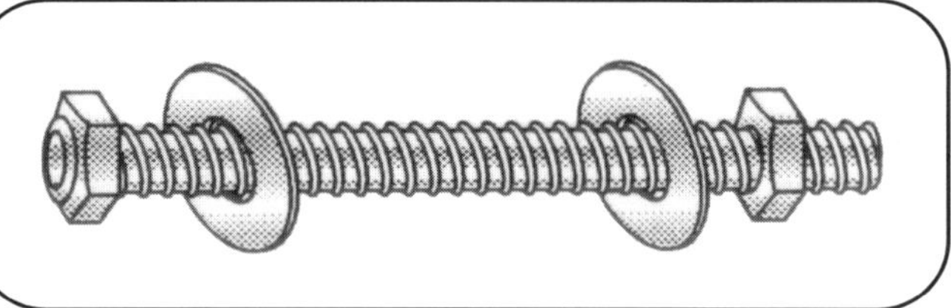

Figure 9.18 Threaded rod with nuts and washers.

Figure 9.19 Sawing the excess rod.

GUYING

Arborists sometimes install guy wires to support trees. **Guying** of trees tends to be permanent because the tree will probably not produce the roots or wood necessary to support itself once it has been guyed. The tree, in fact, will be dependent on this mechanical

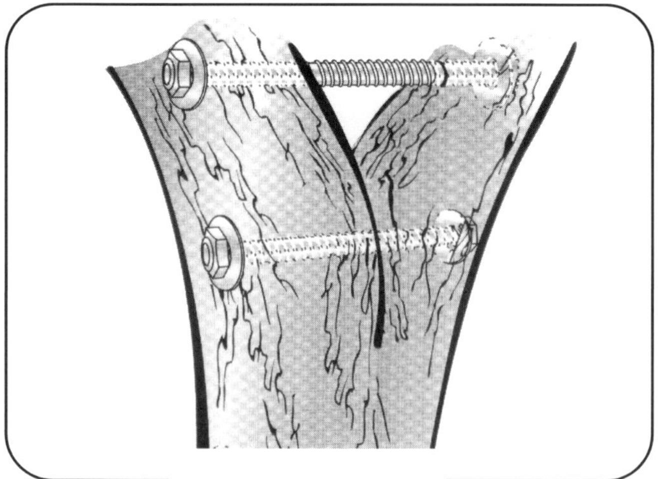

Figure 9.20 Multiple bracing rods are sometimes required. It is best to stagger the rods and avoid close spacing and longitudinal alignment.

support. When considering the installation of guy wires, arborists should carefully assess the risk potential of the tree on that particular site. If the objectives cannot be met by the additional support, other means of mitigating the potential hazards should be considered.

Guy wires can be attached to trees using the same methods for installing cables. As with cable installation, hardware should be installed to be in direct line with guy wire. When installing tree-to-ground guys, the ground anchor must be sufficient to support the tree. Guys should be attached to the tree at a height at or above the midpoint. The ground anchor(s) should be placed no closer to the trunk than two-thirds the distance from the ground to the lowest attachment in the tree. If the tree is to be guyed to another tree, the support tree must be checked to ensure structural integrity and the ability to support the second tree. Guys should be attached to the tree to be supported above the midpoint, preferably at about two-thirds the height of the tree.

A significant consideration in the installation of guying systems is public safety. This involves the additional support of the trees to be guyed, as well as safety concerns related to the guy wires themselves. Tree-to-tree guys should be installed high enough so as not to present a hazard to pedestrians. Tree-to-ground guys should be clearly marked and protected from inadvertent contact.

Figure 9.21 A lightning protection system installed in a tree.

LIGHTNING PROTECTION FOR TREES

A bolt of lightning can destroy a tree in less than a second. Trees can be blown apart, stripped of their bark, or set on fire by lightning strikes. People or animals standing under a tree can be killed when lightning strikes the tree.

It has been observed over the years that some trees are more likely to be struck than others. Trees that stand alone in open landscapes, the tallest trees in an area, or trees standing on a hill have a higher probability of being struck by lightning.

Many circumstances warrant protecting a tree from lightning. Historic trees, trees of great economic value, or large trees within 10 feet of a structure are all candidates for lightning protection. Trees on a golf course or in a park where people seek refuge during a storm should be equipped with lightning protection systems.

A lightning protection system consists of a series of copper conductors that extend from the top of the tree, down the main branches and trunk, and out beyond the tree underground where they are grounded (Figure 9.21). All lightning protection hardware used in trees should be approved by the National Fire Protection Association or the Lightning Protection Institute.

The uppermost point of the system is the **air terminal**. Copper or copper bronze points are manufactured for this purpose but are not required for a system. They are usually attached to the tree using barbed copper nails. The air terminals should be located near the top of any major leaders or on major branches in a tree with a broad crown. A large tree would probably need one central air terminal and three to five secondary terminals. It is not necessary for the terminals to extend beyond the branch tips.

The **main conductor** is a copper cable connected to the primary air terminal and running down a main branch and the trunk. The cable is attached to the tree at regular intervals (usually 3 feet) using approved conductor attachments. Secondary conductor cables connect to secondary terminals and tie into the main conductor on the trunk. The secondary conductors are also attached to the tree at approximately 3-foot intervals. Conductors should be attached to allow enough slack for tree sway. If there are support cables in the tree, they should be connected to the lightning protection system. The conductor must be properly grounded. The conductor cable should extend out from the tree and be buried to a depth of 8 to 12 inches below the soil surface. The cable is attached to a copper-clad or copper-weld **ground rod** that is 8 to 10 feet long and ½ inch in diameter. The ground rod should be driven into the soil to a depth of 10 feet.

In dry, sandy, or rocky soils, it is recommended that the cable be forked to several ground rods (Figure 9.22). If it is not possible to drive the ground rod to a depth of 10 feet, it should be cut below ground level. The cut piece should be driven into the ground at least 6 feet beyond and connected to the first by cable. Grounding resistance of 50 ohms or less is considered preferable, although resistance will vary with soil type and moisture.

If two or more trees in proximity are protected, the grounding cables should be connected. In groups of trees, only the major trees need to be protected. They will provide protection for the smaller trees. Ground rods should not be installed near underground utilities or storage tanks. The grounding conductor must not pass over the top of utilities or tanks.

As with cables and braces, lightning protection systems should be inspected annually. When the tree has grown significantly past the air terminals, the conductor cables and terminals should be extended.

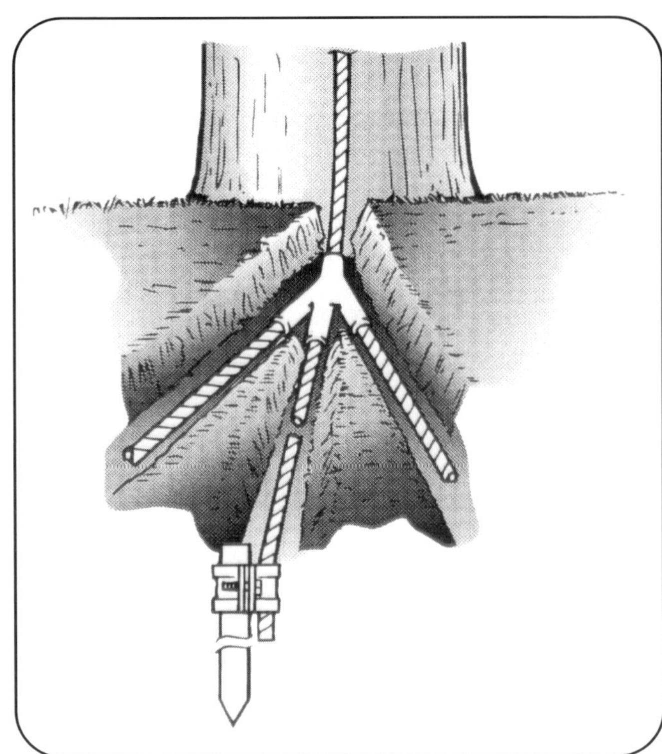

Figure 9.22 In dry, sandy, or rocky soils, the cable should be forked to multiple ground rods.

Eventually the tree may grow enough in girth to envelop the conductors. The system will still be functional as long as there is no break in the conductor. Annual inspections should include a check of all splices and connections to ensure a continuous flow of current.

Chapter 9 Workbook

1. Name three instances in which cabling might be called for in a tree.
   a.
   b.
   c.
2. Name three types of cable systems.
   a.
   b.
   c.
3. Before installing cables in a tree, the tree should be ________________, if necessary.
4. As a general rule, cables should be installed _______ the distance from the crotch to the top of the tree, as long as the wood is solid and large enough to install the hardware.
5. Limbs may be brought closer together while installing the cable so that, when released, the cable will be just _______.
6. Dead-end hardware such as ________ ________ or ________ ________ can be used as anchors in smaller branches that are not decayed.
7. In decayed wood, or limbs greater than 8 inches in diameter, _____ _____ or ____________ _____ _____ _____ _____ _____ must be used to attach cables.
8. True/False—When more than one cable is installed on the same limb, the hardware should be spaced at least as far apart as the diameter of the limb.
9. The two types of cables most commonly used in trees are ___________ __________ ___________ ________ and ______ _______ ______________.
10. A ____________ is used to form the eye at the end of a cable.
11. EHS cable should be attached to hardware using ________ ________ ________.
12. True/False—The installation of metal cables, if done properly, will not wound the tree.
13. The two types of rods commonly used to brace a tree are _____________ ____________ _____ and _____ ____________ _____.
14. When bracing a tree with a machine-threaded rod, the pre-drilled hole should be approximately $\frac{1}{16}$ to $\frac{1}{8}$ inch __________ than the diameter of the rod.
15. Washers attached to the ends of bracing rods should be ____________ or ___________ in shape.
16. Name three circumstances in which lightning protection for trees might be recommended.
    a.
    b.
    c.
17. All lightning protection hardware used in trees should be approved by the _____________ _____ __________________ ___________________or _________________ ___________________ __________________.
18. The uppermost points of a lightning protection system are called the ____ _______________.
19. The main conductor in a lightning protection system is a cable made of ________.
20. The conductors in a lightning protection system are attached to the tree at approximately _____-foot intervals.

21. True/False—The air terminals must extend higher than the tips of the branches for a lightning protection system to be effective.
22. True/False—If a tree grows around the standard down conductor, the lightning protection system will not function properly.
23. The ground rod of a lightning protection system should be driven to a depth of ______ ___________, if possible.
24. True/False—Lightning protection systems need not be grounded in sandy soils.
25. True/False—The ground cable in a lightning protection system should never be forked.

CHALLENGE QUESTIONS

1. What are the differences between 7-strand, common-grade galvanized cable and extra-high-strength cable? What are the advantages and disadvantages of each?

2. Ideally, a cable should be installed in a tree perpendicular to the line that bisects the angle of the crotch. Diagram this installation and discuss how it affects hardware installation.

3. Why is it important to balance the relative strengths of the cabling and bracing materials used to provide support in a tree?

4. Draw a diagram of a lightning protection system labeling the air terminals, conductors, ground cable, and ground rod.

5. Discuss lightning protection for a grove of several large and small trees.

SAMPLE TEST QUESTIONS

1. Which of the following must be true in order for a lightning protection system to be effective?
   a. the tree cannot have grown around the conductors
   b. the conductor must not be forked
   c. the conductor must be properly grounded
   d. all of the above

2. The purpose of a lightning protection system is to
   a. reduce the voltage of the strike
   b. prevent the tree from being struck
   c. conduct the electrical charge into the soil away from the tree
   d. all of the above

3. If two bracing rods are installed to support a tree crotch, they should be placed
   a. no more than 4 inches apart
   b. in vertical alignment, one above the other
   c. staggered, and no closer together than the diameter of the trunk
   d. never use more than one brace rod per crotch

4. If a lag hook is installed into decayed wood
   a. the decay will likely spread along the lag
   b. the holding power will be reduced
   c. the decay may be detected when drilling
   d. all of the above

5. When cabling a multi-stemmed tree, extra support can be added to the system by
   a. cabling the limbs together in triangular combinations
   b. using double cables on each lag
   c. installing the cables low in the tree with turnbuckles that can be tightened from the ground
   d. all of the above

Other Sources of Information

(See pages v–vi for complete bibliographic information.)

ANSI A300. *Standard Practices for Tree, Shrub, and Other Woody Plant Maintenance, Part 3: Support Systems–Cabling, Bracing, and Guying.*

Harris et al., 1999. *Arboriculture: Integrated Management of Landscape Trees, Shrubs, and Vines.*

CHAPTER 10

DIAGNOSIS AND PLANT DISORDERS

CHAPTER 10 DIAGNOSIS AND PLANT DISORDERS

CHAPTER 10
DIAGNOSIS AND PLANT DISORDERS

Objectives

1. Be able to distinguish between plant problems caused by living and nonliving disorders.
2. Understand the principles of a systematic approach to the plant diagnostic process.
3. Explain how stress can weaken a tree, predisposing it to additional problems.
4. Learn to recognize the signs and symptoms of tree disorders.
5. Learn the various physiological disorders and injuries that can affect trees, and learn what treatments are appropriate.
6. Become familiar with various types of insect and disease problems and their impact on trees.

Key Terms

abiotic disorders	chronic	leaf blotch	powdery mildew	symptom
acute	complex	leaf spot	rust	systemic
allelopathy	defoliation	necrosis	scorch	vascular discoloration
biotic disorders	dieback	nematodes	sign	vector
blight	gall	pathogen	skeletonized	wilt
canker	gummosis	physiological disorder	stress	witch's broom
chlorosis	infectious		stunting	

INTRODUCTION

Diagnosing plant problems requires a combination of knowledge, experience, and keen observation. Most often it is not simply a case of identifying an insect or disease. Trees that die or decline usually are suffering from a combination of stress factors, and insects or diseases are often secondary, attacking the weakened tree. It takes some detective work to piece together all of the clues (Figure 10.1). Then, arriving at a diagnosis is often only the first step. The important second step is to make appropriate recommendations.

Figure 10.1 Diagnosing tree problems requires detective work. Gather all of the clues before making a recommendation.

GENERAL DIAGNOSIS

The biggest challenge facing the arborist in diagnosing tree problems is the lack of available information. Obviously, the tree cannot describe its symptoms; thus, it is up to the diagnostician to uncover the answers. The history of the tree and its environment is of utmost importance. An arborist must learn to ask key questions about the site and the tree over the years. How long has the problem been going on?

What were the early symptoms? Have there been any construction, excavation, or chemical treatments in the area? Sometimes asking the same question in different ways helps the client reveal more information. Keep in mind that the information received may not be entirely accurate. It is not unusual to be told that a tree seemingly died overnight when it had actually been declining for years. Often the arborist isn't called in to treat the tree until it is already dying or dead.

It is helpful to categorize plant health problems into two major groups, living and nonliving. Living agents (**biotic**) include plant pathogens such as fungi, bacteria, viruses, and **nematodes**, as well as insect pests, mites, and other animals. These are considered to be **infectious** because they can spread from one plant to the next. Nonliving agents, also called **abiotic disorders**, are noninfectious and include environmental problems such as temperature and moisture extremes, mechanical injuries, chemical injuries, mineral deficiencies, and many others.

Diagnosis is a systematic process that involves information gathering, keen observation, and logical analysis. Unfortunately, there is a tendency to shortcut the process, ending up with an incomplete diagnosis or a misdiagnosis. Correct diagnosis of tree problems requires careful examination of the situation and systematic elimination of possibilities by following a few important steps.

1. Accurately identify the plant. Many insects and diseases are specific to certain host trees. In addition, certain species of plants are prone or sensitive to specific abiotic disorders. Knowing the identity of the plant can quickly limit the number of suspected causes. Inherent in identification, though, is the knowledge and recognition of what is normal. For example, a common complaint of home owners is that their pine trees are dropping needles. Knowing that it is normal for pines to shed their inside needles, an arborist will quickly determine whether it is the inside, older needles or the new-growth needles that are dropping.

2. Look for a pattern of abnormality. Differences in the patterns of damage between biotic and abiotic disorders are often key in diagnosing plant problems. Again, an arborist must first know what is normal for that plant. It may be helpful to compare the affected tree with other plants on the site, especially those of the same species. Differences in color or growth may present clues as to the source of the problem. Nonuniform damage patterns can be indicative of living factors such as insects or pathogens—problems caused by living agents rarely follow straight lines. In contrast, if there is a clear border between affected and unaffected areas, the cause is more likely to be abiotic. Uniform damage over a large area (perhaps several plant species) usually indicates nonliving factors such as physical injury, chemicals, or weather, to name a few.

3. Carefully examine the site. Check the contour of the land and note the structures present. If affected plants are restricted to a walkway, road, or fence, the disorder could be a result of wood preservatives, de-icing salts, or other harsh chemicals. The history of the property may reveal situations such as grade changes, excavations, herbicide applications, or gas leaks. Examine the area or community where the tree is located. How do plants of the same species look? Is the area a new community with imported plants not native to the area? Is it an older community where age and overcrowding can be limiting factors? Or is it a new housing division in an old woodlot? The general area can often set the stage for certain types of problems.

4. Note the color, size, and thickness of the foliage. The foliage is invariably the first part of the plant that most people notice. Changes and irregularities in the leaves tend to be easy to spot, and symptoms are often obvious. Many disorders become apparent in the foliage. Dead leaves at the top of the tree are often the result of mechanical or environmental root stress. Twisted or curled leaves may indicate viral infection, insect feeding, or exposure to herbicides. Early fall color can be a symptom of stress from girdling roots or other root-related problems. The list can go on almost indefinitely. The tricky part is that many disorders can have similar effects on the leaves. In addition, the cause of the problem often begins elsewhere in the tree.

5. Check the trunk and branches. More clues and symptoms can be found on the trunk, branches, and twigs. Examine the trunk thoroughly for wounds because they provide entrances for canker and wood-rotting organisms and can disrupt the transport of materials between roots and leaves. Small holes may indicate the presence of borers or bark beetles. Profuse development of watersprouts can indicate a variety of stresses. A good arborist is thorough and systematic in this investigative process, being careful not to miss any clues.

6. Examine the roots and root collar. The root system is the most frequently overlooked portion of the tree in the diagnostic process. This is unfortunate because the belowground part of the plant is often the "root" of the problem. Stress caused by poor site conditions, bad planting practices, and humanmade site changes or management practices probably account for most of the initial plant health problems in urban sites. Take the time to examine the site and soil, the root collar, and the rest of the root zone. Analyze the soil and moisture conditions. Check for signs of decay in the major roots and collar. Note the color of the roots: Healthy, fibrous roots are generally white and fleshy. Brown roots may indicate dry soil conditions or the presence of toxic chemicals. Black roots often reflect overly wet soil or the presence of root-rotting organisms.

When diagnosing a tree problem, try to systematically rule out certain possibilities. The majority of tree health problems are not caused by insects, mites, fungi, or bacteria. Rather, 70 to 90 percent of all plant problems result from adverse cultural and environmental conditions such as soil compaction, drought, moisture fluctuations, temperature extremes, mechanical injuries, or poor species selection. Rarely is poor plant health the result of one simple factor; the cause is usually a combination or **complex** of nonliving stresses and living contributors.

Diagnostic ability is a cumulative skill. The best diagnosticians learn every time they look at trees. Learn to gather as much information as possible before committing to a diagnosis. Keep an open mind. As an example, the arborist who finds wood decay, cankers, or borers and dwells exclusively on the fungus or insects rather than the reason for their presence may be concentrating more on the symptoms than the cause. Many insects and fungi are secondary, and only become a problem on stressed trees.

If, after going through the diagnostic process, the source of the problem is still unknown or not clear, it may be advisable to seek help. One of the biggest mistakes one can make is to guess at the diagnosis and recommend treatment based on that guess. Part of the learning process is to recognize when additional expertise is required and to know where to go to obtain it.

SYMPTOMS AND SIGNS

In diagnosing tree problems, arborists must look for **symptoms** and **signs** to determine the cause. Symptoms are effects of the causal agents that are apparent on the tree, or how the tree responds to the disorder. Examples of symptoms include chlorosis, wilting, and leaf scorch. Rarely can a problem be diagnosed by a single symptom. Wilting, for example, can be the result of drought, root problems, or various fungal or bacterial organisms. Signs are direct indications of primary or secondary causal agents, or something "left behind" by the agent. Signs might include conks or fruiting bodies of fungi, insect frass, emergence holes, or discarded "skins." Some of the more common tree ailment symptoms are defined below (Figure 10.2).

Leaf spot—spots of dead tissue on the foliage; the size, shape, and color may vary with the causal agent but are usually limited to a small portion of the leaf

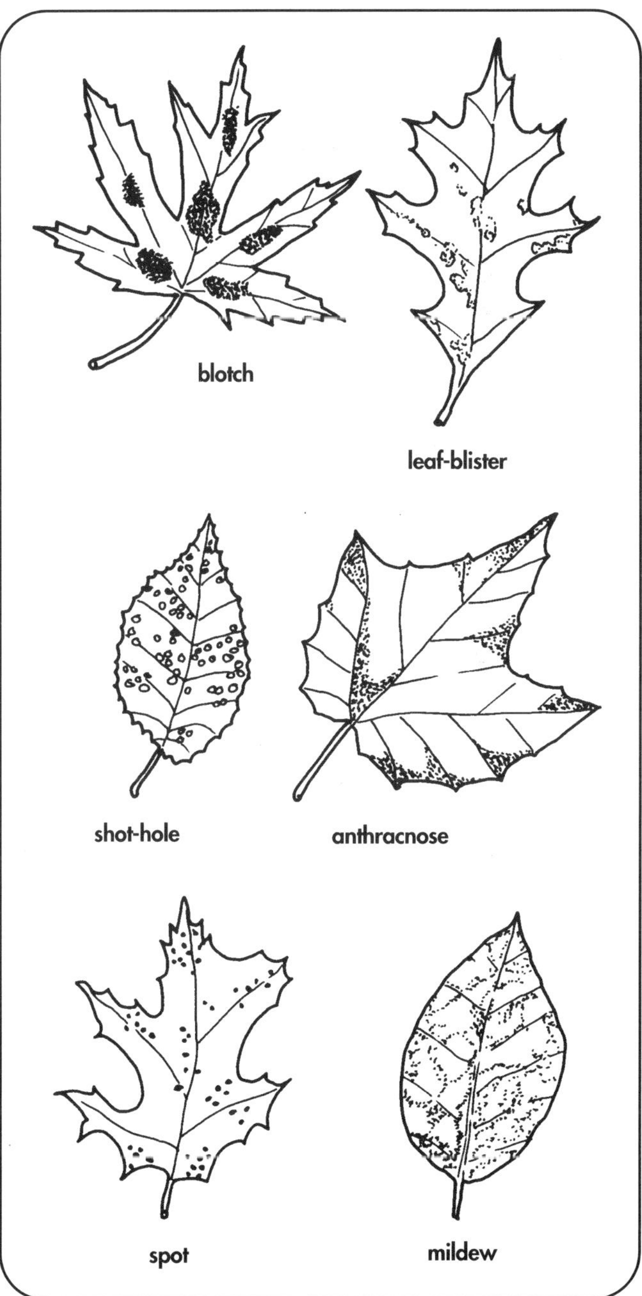

Figure 10.2 Patterns of damage on the foliage can indicate the possible cause of a tree's problem.

Leaf blotch—dead areas of tissue on the foliage, irregular in shape and larger than leaf spots

Blight—dieback of leaves and twigs on major portions of a tree, especially the young, growing tissues

Scorch—browning and death of indefinite areas along leaf margins and between veins

Wilt—drooping of leaves or shoots, often due to lack of water in the tree

Canker—localized dead stem tissue, often shrunken and discolored

Stunting—abnormally small plant growth

Gummosis—exudation of sap or gum from wounds or other openings in the bark

Rust—orange or reddish brown pustules on leaves, or galls and cankers on stems caused by certain fungi

Gall—swollen plant tissue, or irregular plant growth that may be induced by insects, fungi, bacteria, or nematodes

Chlorosis—yellowing of normally green tissues due to a lack of chlorophyll

Necrosis—death of tissue

Dieback—dead portions in the tree

Powdery mildew—white or grayish fungal growth on the surface of plant tissues, usually leaves

Vascular discoloration—darkening of the wood's vascular elements, often along growth rings

Witch's broom—abnormal development of multiple secondary shoots, forming a broomlike effect (Figure 10.3)

Figure 10.3 Witch's broom on hackberry (*Celtis occidentalis*).

Many symptoms observed in declining trees are nonspecific. They may not lead to a direct diagnosis because there can be many causes. Nonspecific symptoms must be analyzed in combination with specific information about the tree and the site in forming a diagnosis (Table 10.1).

TREE STRESS

The basic factors that promote plant health include sufficient water, optimal temperature and light, and a proper balance of minerals. Too much or too little of any of these elements can cause stress. **Stress** is a term used to describe any condition that causes a decline in tree health.

Stress may be classified into two broad categories: **acute** and **chronic**. Acute stress, which can be caused by such things as improper pesticide applications or untimely frosts or freezes, occurs suddenly and causes almost immediate damage. Chronic stress takes a longer time to affect plant health and may be the result of such things as mineral nutrient deficiency, improper soil pH, long-term weather changes, incorrect light intensity, or other factors.

Early signs of stress might include reduced growth, abnormal foliage color, vigorous watersprouting, or premature leaf drop. The most common causes of tree stress are site and/or environment related. If a tree is not well suited for the site in which it has been planted, it is more likely to become stressed. Frequent stress problems include excess or inadequate water, extreme cold, soil conditions such as compaction, or mechanical injury from construction damage.

Stress is not necessarily an irreversible condition. If the stress factor(s) can be identified, it is sometimes possible to alleviate the condition and improve the health of the tree. However, if only the secondary problems are identified and treated, it is likely that the tree will continue to decline.

ABIOTIC DISORDERS

Abiotic disorders affect the normal growth and development of a tree. They may disrupt water and mineral uptake or other vital functions, which sometimes creates a domino effect of physiological dysfunction within the tree. Many agents of abiotic disorders are primary stress factors that can become the first stage in a spiral of decline. Each successive stress that follows adds to the decline, often further reducing the possibility of reversing the trend.

Signs of abiotic disorders are not obvious and may be difficult to recognize. Learning the history of the tree and site often provides the biggest clues.

Table 10.1 Diagnostic guide for landscape plants exhibiting nonspecific symptoms.

SYMPTOMS	POSSIBLE CAUSES
1. Brown or scorched leaves; progressive dieback of branches	A) Poor root health from poor drainage, excessive soil dryness, excessive fertilizer, compaction, and poor water penetration into soils or girdling roots B) Specific nutrient toxicities or imbalances C) Excessive heat or light reflected onto leaves from driveway or buildings D) Pesticide or mechanical injury E) Air pollution. F) Winter drying G) Vascular fungal or bacterial infection
2. Leaf spots, blotches, blemishes, blisters, or scabby spots	A) Excessive soil dryness coupled with high temperatures B) Frost injury C) Chemical spray injury D) Fungal or bacterial infections E) Herbicide injury F) Insect damage
3. Foliage yellow-green	A) Nutrient deficiencies B) Poor root health due to compacted soil, poor drainage, or girdling roots C) Winter drying D) Root or crown injury E) Air pollution F) Soil pH lower than 5.0 or higher than 8.0 G) Herbicide injury H) Mites or scale
4. Foliage of one branch dying	A) Fungal canker B) Injury C) Insect damage D) Winter damage E) Chemical spray injury
5. Leaf drop	A) Poor root health from poor drainage, excessive dryness, excessive fertilizer, compacted soil, or girdling roots B) Heat and drought stress C) Insect infestation D) Herbicide injury
6. Wilting or drooping of foliage	A) Poor root health from poor drainage, excessive dryness, excessive fertilizer or other soluble salts in the soil, compacted soil, or overwatering B) Toxic chemical poured into soil C) Fungal or bacterial infection of vascular system D) Fungal cankers E) Insect infestation
7. Leaves with tiny yellow speckling or yellow banding of needles	A) Mite infestation B) Air pollution C) Insect infestation D) Fungal or bacterial infection
8. Deformed or misshapen leaves	A) Herbicide injury B) Late frost or freeze C) Insect infestation D) Anthracnose E) Spray injury

To more correctly determine causes of landscape problems, compile lists or sets of symptoms. Look for specific symptoms associated with these nonspecific symptoms.

Knowing about temperature or moisture extremes in recent years, for example, can help. Unfortunately for the diagnostician, however, it can take a long time for the tree to react to the primary stress factors. Because of this, the link between cause and effect is often lost.

Keep in mind that most trees in decline are suffering from a complex of problems. A mature oak (*Quercus* spp.) tree may have suffered stress induced by construction damage to the root zone. Roots are lost or damaged, water and mineral uptake decreases, and the tree declines over a period of ten years. *Armillaria*, an opportunistic root-rot fungus, becomes well established until the tree eventually fails in a windstorm. How many diagnosticians will trace the failure back to the initial construction-related stress? Arborists must understand tree physiology and must learn to look at the whole picture if they are to be good diagnosticians.

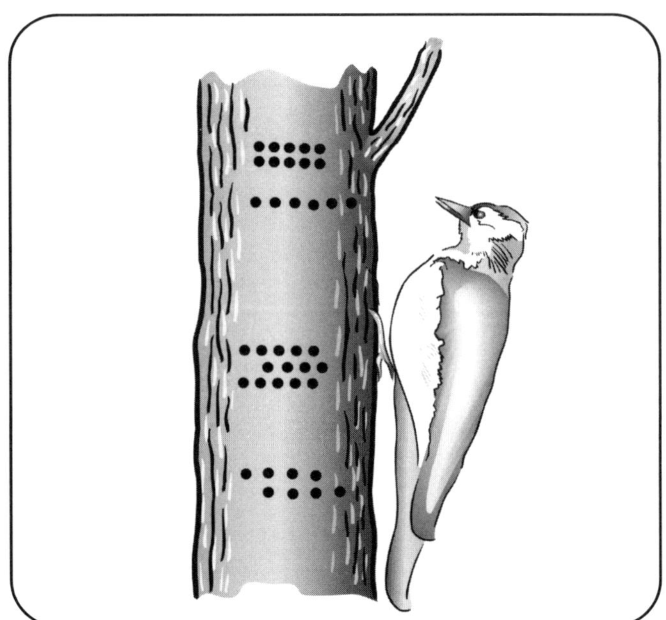

Figure 10.4 Sapsucker damage. Do not mistake sapsucker damage for borer exit holes, which would not exhibit this linear pattern.

Soil and Site Problems

A common mistake in diagnosis is to carefully examine the trunk and crown of the tree while overlooking the root condition. Root-related problems are often difficult to diagnose due to their limited accessibility. Often symptoms observed in the upper portions of a tree result from poor root health. Hundreds of things can cause root health problems, and the symptoms may be similar for many.

If a transplanted tree declines or dies within the first year, the most likely cause is either a lack or an excess of water (which may be related to planting depth). If the tree's root ball is too small, or a saturated condition restricts new root formation or water absorption, the tree will suffer not only from the lack of water, but also from mineral deficiencies.

Another common site problem is compacted soil. Without sufficient pore space, water and particularly oxygen, will be limited. Biological activity in the soil that is beneficial to trees can be inhibited, and root growth and water absorption will decrease, causing the tree to decline or die.

Physical and Mechanical Injuries

It is important to understand the difference between "physical" injuries and **physiological disorders.** Physical or mechanical injuries are relatively sudden, such as fire injury, frost, animal feeding, lightning, vandalism, and lawn mower damage (Figure 10.4). Physiological disorders usually occur over a long period of time.

Often the full extent of damage due to physical injuries cannot be immediately assessed. In some cases, initial treatment should be limited until the extent of tree damage can be determined. For example, with lightning injury, it may be a year or more before the effects of electrical current passing through a tree's tissues are exhibited. Sometimes the causes of physical damage can be eliminated to prevent injury to the tree. For example, where lawn mower injury is a problem, the turf should be removed from the area around the tree and replaced with mulch.

Temperature Extremes

Extremes in temperature, either high or low, can be damaging or even lethal to trees. Frost damage can cause blackened or distorted foliage. A hard freeze can kill plants that aren't hardy in a given climate, especially if the roots are relatively unprotected. If the roots are damaged, other symptoms may appear later. Trees with thin bark may experience sunscald or frost cracks due to rapid warming and cooling on the south or southwest sides of the trunk. Often these cracks begin at old wounds. High temperatures, reflected heat (such as from pavement or buildings), and drought can combine to stress trees as well. Marginal leaf scorch is an early symptom; wilting, reduced growth, and general decline may follow.

Competition and Allelopathy

Trees can decline as a result of competition with other trees and plants. The most commonly seen example is competition with other trees for sunlight. Trees growing under the canopy of larger trees may exhibit dieback of lower branches, stem curvature, and reduced growth. Another form of competition takes place below the soil line. Trees may compete with lawns and other plantings for available soil nutrients, water, and growing space.

Allelopathy is the chemical inhibition of growth and development of one plant by another. Many trees and other plants produce chemical substances that affect the growth of other plants. Allelopathic chemicals may be exuded directly from the plant or released indirectly through decomposition. These chemicals may inhibit growth, seed germination, flowering, or fruiting of nearby plants. In most cases, the inhibition is minor and not easily diagnosed. Stressed plants are more susceptible to allelopathic effects than nonstressed plants are.

A few trees are considered highly allelopathic: walnuts (*Juglans* spp.), sugar maple *(Acer saccharum)*, black locust *(Robinia pseudoacacia)*, cherry (*Prunus* spp.), hackberry (*Celtis* spp.), some eucalyptus (*Eucalyptus* spp.), and sassafras *(Sassafras albidum)*. The classic example is black walnut, *Juglans nigra*. Black walnut produces a toxin called juglone. This substance restricts root growth of other plants in the area. Many plants will not grow at all in the vicinity of a black walnut tree (Figure 10.5).

Young trees are most seriously affected by allelopathic chemicals. Tall fescue and other grasses can markedly retard the growth of young trees unless the turf is kept at least 2 feet from the trunk. Allelopathy is usually not a problem once a tree has an extensive root system.

Pollution Damage

Pollution (air, water, and soil) damage to trees may be divided into two categories. Acute toxicity results from exposure to high concentrations over a relatively short period of time. Chronic injury is due to long exposures, usually in much lower concentrations. The extent of plant injury is related to the type of pollutant and the rate and duration of exposure. Other factors that affect pollution damage are wind, humidity, soil grade and type, precipitation levels, and general condition of the tree. Stressed trees may show more susceptibility to pollution injury than unstressed trees do. Pollution damage is often difficult to diagnose unless the source of the pollutant is known. The symptoms may mimic other problems such as insect injury and mineral deficiencies.

The four major airborne pollutants that cause damage to trees are sulfur dioxide, fluoride, ozone, and peroxyacetyl nitrates (PAN). Dust and airborne particles may contain a combination of pollutants. When combined with rain or dew, these particles may become more toxic.

Most of the symptoms of pollution damage will be evident on the foliage. Dieback or necrotic areas may appear on the leaf tips, along the margins, or between the veins (Figure 10.6). The leaves may appear whitish or silvery. There may be some stippling or spots on the leaf surface. Conifer needles may have chlorotic banding or tip dieback. Reduced growth is common.

If pollution injury is suspected, the most obvious solution is to eliminate the pollution at the source. However, this is usually not a practical option. If the exposure has been acute, it may be beneficial to thoroughly wash the foliage with a water spray on a regular basis when pollutants are present. In sites where chronic pollution is a problem, it is best to select plants that are tolerant of the pollution.

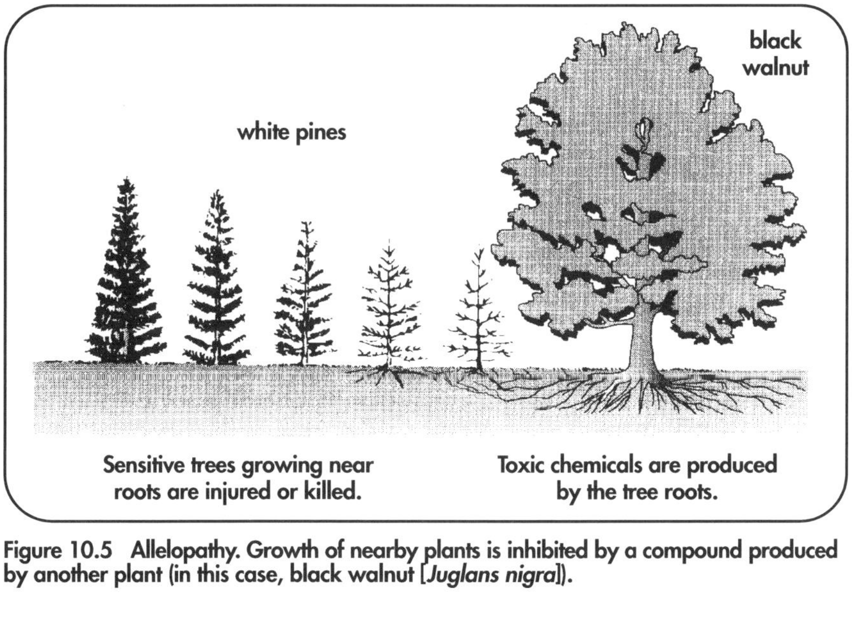

Figure 10.5 Allelopathy. Growth of nearby plants is inhibited by a compound produced by another plant (in this case, black walnut [*Juglans nigra*]).

Chemical Injury

Although any number of chemicals can kill or injure a tree, herbicides are the most frequent culprits. Because herbicides are formulated to kill plants, they are the most phytotoxic of the pesticides used in the landscape. The growth regulator herbicides, such as 2,4-D and dicamba (used to kill broadleaf weeds in the lawn), may harm nontarget broadleaf plants such as trees and shrubs. These materials are **systemic**, which means they move throughout the plant.

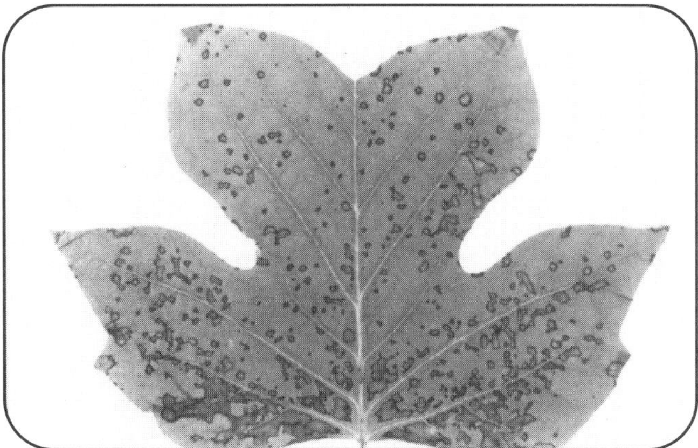

Figure 10.6 Sulfur dioxide injury on tuliptree (*Liriodendron tulipifera*).

Exposure may be a result of drift, accidental spraying, or movement into the soil.

If a tree has been exposed to these materials, the leaves will often begin to curl and cup, and the shoot tips may become twisted. New foliage will show varying degrees of parallel venation. The plant may appear wilted. The foliage may become yellowish before dying and falling off (Figure 10.7). The plants often recover unless exposure is severe. Exposure to other herbicides may cause trees to exhibit similar symptoms. Other symptoms include veinal or interveinal chlorosis, marginal chlorosis, and leaf-fall. Certain herbicides (nonselective herbicides or soil sterilants) have greater potential to cause serious damage than most growth regulator herbicides, and caution is in order whenever these are used near trees or landscape plants.

To minimize the chances of accidental herbicide injury, never use the same spray equipment to apply herbicides that is used for fungicides or insecticides. Spray only on cool, calm days and avoid drift. Using low pressure increases droplet size and reduces drift. Apply the herbicide to target plants only, and always check the product label for possible phytotoxicity to the target or surrounding plants. Always use great caution when applying herbicides within the root zone of trees. If they must be used within the root zone, be sure the product is labeled for such use.

BIOTIC DISORDERS

Insects and Other Pests

At any given time, there may be several species of insects present on a tree, although only a few, if any, may be harmful. However, almost every species of tree has at least one insect pest that can cause some

Figure 10.7 Herbicide damage on dogwood (*Cornus kousa*).

problems. Some insects (Japanese beetle, aphids, some scales) will feed on a variety of host plants, while others are specific to certain hosts. Many insects are predators or parasites of harmful insects. Insects have complex life cycles; one stage may cause problems, while the next does not. Knowledge of the pests' life cycles is important in identification and treatment. An arborist must be able to identify both harmful and beneficial insects, and know the extent of damage the plant can tolerate before any control is necessary. Reference books and diagnostic labs should be used when needed (Figure 10.8).

Most insect damage to trees is the result of either feeding or egg-laying activity. Periodical cicada can destroy small twigs by ovipositing (laying eggs) in long lines under the bark. Because cicadas may be present in such large numbers, this wounding can be a major problem during years when major cicada broods emerge.

Insect feeding damage is characterized by the type of mouthparts the insect has. Insects such as caterpillars, webworms, beetles, and weevils chew on plant parts. Chewing insects eat plant tissue such as leaves, flowers, buds, and twigs. Some insects, such as gypsy moth, Eastern tent caterpillar (Figure

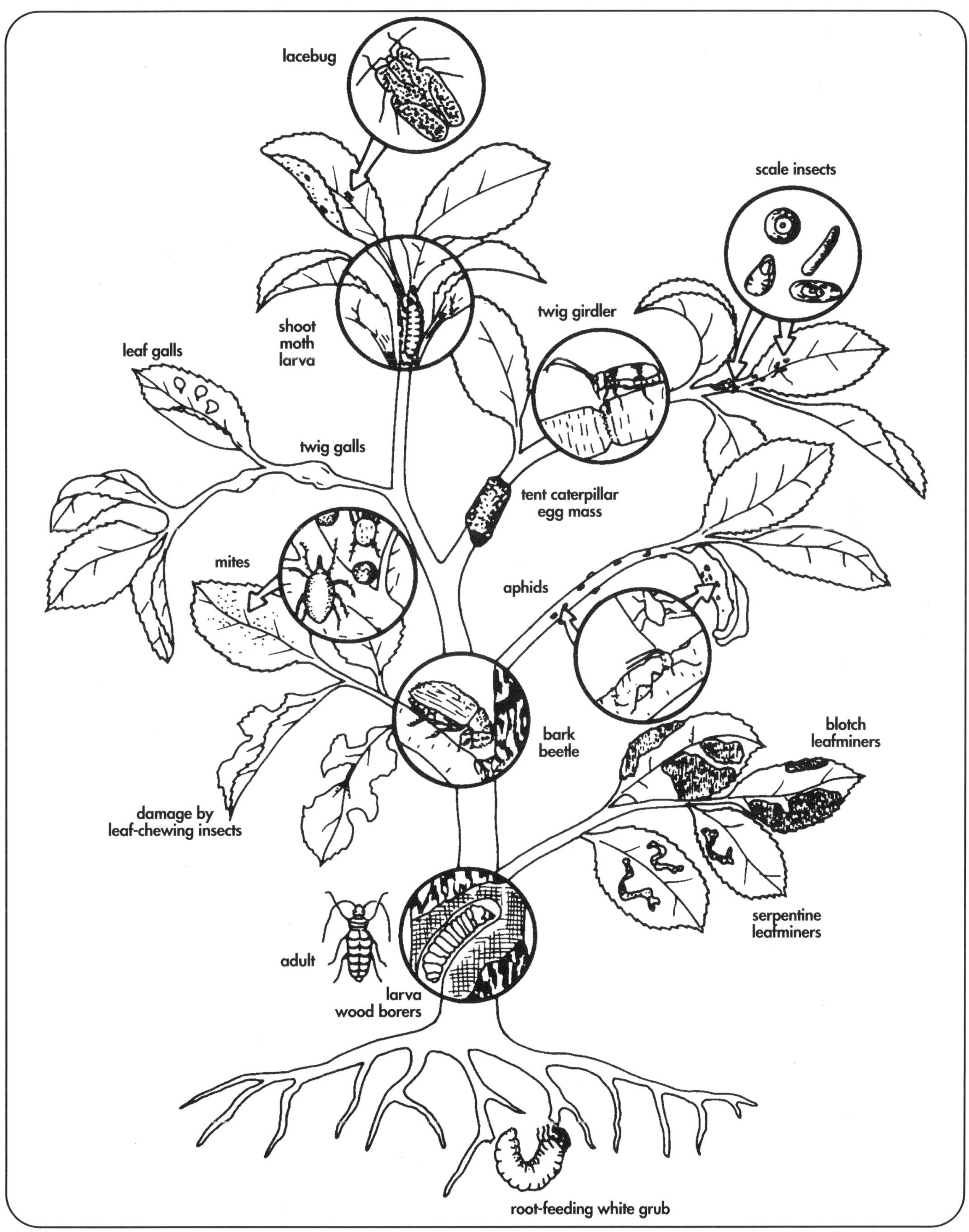

Figure 10.8 Various insects and the injuries they cause.

10.9), and cankerworm eat the entire leaf. Black vine weevil, however, feeds on the leaf margins. Indications of damage by this insect are often seen as uneven or broken margins, or notches on leaves. Other insects, such as Japanese beetle and elm leaf beetle, eat only the interveinal tissue, creating a **skeletonized** leaf (Figure 10.10). Leafminers feed between the leaf surfaces, hollowing out tunnels inside the leaves (Figure 10.11).

Borers are chewing insect larvae that tunnel under the bark and often into the wood of trees (Figure 10.12). Because each kind has its own style and tunnel pattern, borers may be identified by their work even after they have left the scene. Trees infested with borers typically show a thinness of crown and a decline in vigor. Conclusive symptoms are small emergence holes in the trunk or branches with frass (semi-digested wood). Borers eat the inner bark, phloem and cambium, or xylem, thereby destroying the tree's ability to transport water and nutrients between the roots and crown of the tree. Some borers, like the Asian longhorned beetle, tunnel into the wood of the plant and cause structural damage.

Other insects feed by piercing and sucking. Aphids/adelgids, scales, leafhoppers, and true bugs feed by piercing the plant cells and sucking out the contents (Figures 10.13 through 10.16). Symptoms of this type of feeding include chlorosis, stippling, drooping, and distortion. A few of these insects can also cause phytotoxic effects as a result of the chemicals they secrete into the plant while feeding. Certain scales can cause serious decline in trees, partially because they tend to go undetected for years. Generally, however, most sap- or leaf-feeding insects do not

Figure 10.9 Eastern tent caterpillar.

Figure 10.10 Japanese beetle on linden (*Tilia* spp.).

Figure 10.11 Birch leafminer.

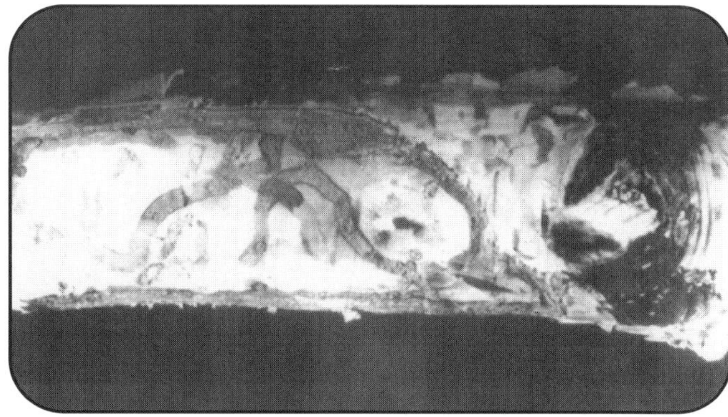

Figure 10.12 Borer damage.

Figure 10.13 Aphids on crabapple (*Malus* spp.).

Figure 10.14 Pine needle scale.

Figure 10.15 Oystershell scale.

Figure 10.16 Cottony maple scale on silver maple (*Acer saccharinum*).

kill trees outright, but they can be an additional stress factor. Also, insect drippings or droppings can create a nuisance (Figure 10.17).

Mites are close relatives of insects but are actually arachnids, as are spiders and ticks. They are very tiny, and a hand lens is usually required for identification. Spider mites cause stippling or bronzing of the foliage and sometimes leaf drop as a result of their sap-feeding activity (Figure 10.18). Eriophyid mites often cause galls to form on foliage and twigs as a result of feeding or egg laying. These galls are

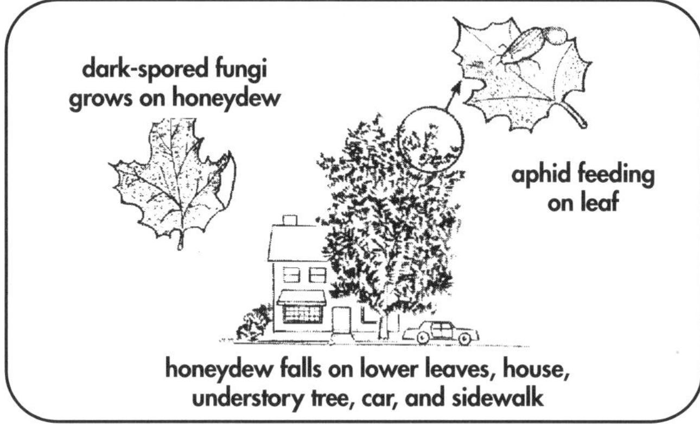

Figure 10.17 Sooty mold grows on honeydew secreted by some sucking insects.

actually plant tissue responses to the insect or mite activity (Figure 10.19).

Some insects are **vectors**, or carriers, of plant diseases. This means that they transmit or spread the **pathogen**, or disease-causing organism, from tree to tree. Dutch elm disease is an example of a fungal pathogen that is often spread by an insect vector, bark beetles (Figure 10.20). Fireblight bacteria can be spread by bees as they collect nectar from flowers. Aphids and leafhoppers can transmit viruses.

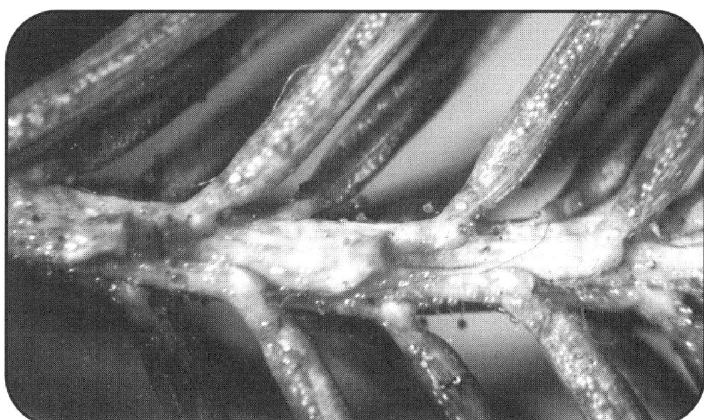

Figure 10.18 Red spider mites on spruce (*Picea* spp.).

Figure 10.19 Ash flower gall.

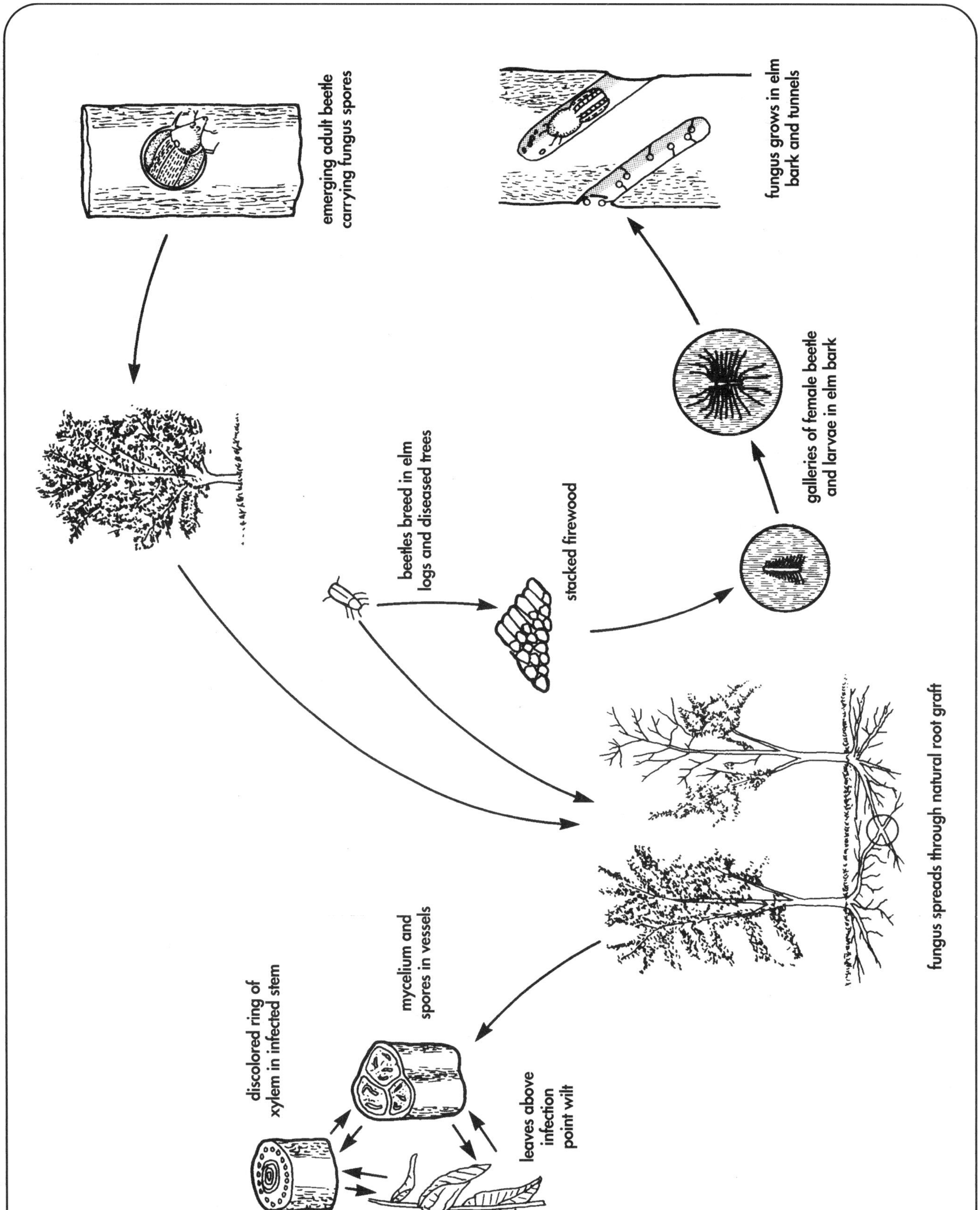

emerging adult beetle carrying fungus spores

fungus grows in elm bark and tunnels

galleries of female beetle and larvae in elm bark

beetles breed in elm logs and diseased trees

stacked firewood

fungus spreads through natural root graft

mycelium and spores in vessels

discolored ring of xylem in infected stem

leaves above infection point wilt

Figure 10.20 Cycle of Dutch elm disease.

Nematodes

Another group of common plant pests are nematodes. Nematodes are microscopic roundworms. Hundreds of species of nematodes have been identified. Not all nematodes are parasites of plants. Most live in the soil and are extremely abundant. They are important promoters of organic decay, a necessary step in the process of natural nutrient cycling. Pathogenic or harmful nematodes usually enter the tree through roots, wounds, stomata, or even directly through plant cells. Nematode feeding can cause swelling, deformation of plant parts, blockage of vascular tissue, and even death of the tree. Some nematodes are vectors (carriers) of pathogens.

Diseases

Four requirements are necessary for a tree disease to become serious: a susceptible host, a pathogenic organism, an environment suitable for disease development, and the right timing. Most pathogens are host specific, meaning they attack a specific plant species or family. A few pathogens attack a broad range of trees.

Often the part of the tree affected is an indication of the severity of the disease. For example, diseases that affect only the foliage may not be a major health problem unless **defoliation** occurs in several consecutive years. However, vascular diseases, those found in the conducting tissues of a tree, such as oak wilt and Dutch elm disease, tend to be fatal.

The vast majority of plant diseases are caused by fungi (Figures 10.21 through 10.23). However, that does not mean that most fungi cause disease. Only a small percentage of fungi cause disease in plants. Many are beneficial. Most trees are susceptible to infection by at least one disease-causing fungus. The severity and extent of the disease depends upon the resistance and vitality of the plant, the virulence of the organism, and the environmental conditions. Most fungal pathogens require moist conditions and specific temperature ranges to initiate infection. Symptoms of fungal diseases vary; some cause leaves to have a dry texture, some form concentric rings, and others produce fruiting organisms that can be used in identification.

Besides fungi, there are many other disease-causing organisms or pathogens. Two common diseases, fireblight and crown gall, are caused by bacteria. Plant tissues infected with bacteria frequently appear water soaked and may have a foul odor. Other diseases such as ringspot and yellowing and some stunt diseases are caused by viruses and phytoplasmas (PLOs).

Figure 10.21 Anthracnose on ash (*Fraxinus* spp.).

Figure 10.22 Apple scab.

Figure 10.23 Powdery mildew on lilac (*Syringa* spp.).

Control of plant diseases may be considered an exercise in prevention. Most fungicides are best used to *prevent* the establishment of disease rather than to *eliminate* existing disease problems. Steps taken to maintain tree health and improve cultural conditions are often the best disease-prevention measures.

GETTING LABORATORY ASSISTANCE

In diagnosing tree health problems, it may be helpful to get laboratory assistance. Diagnostic labs can isolate pathogens, identify pests, and keep arborists informed of potential problems of the season.

Most states have diagnostic facilities available to the public either through universities, the local extension service, or the USDA. If practical, it is better to bring plant samples to the lab and describe the problem to the laboratory technician. Alternatively, include written descriptions with color photographs of the problem.

If the sample must be mailed, it should be freshly cut and sealed in a bag. Wet or moist plant samples mailed in plastic bags arrive at the laboratory mildewy or rotted. Package the sample to minimize damage, and mail early in the week to avoid having the sample spend a hot weekend in the post office. Samples should be taken from representative portions of the plant. If a disease organism is to be isolated, the sample should contain the interface between diseased and healthy tissue. Most important of all, a detailed description of the problem, the host plant, and the surrounding conditions should accompany the sample to the lab. Rarely is the real problem a simple matter of isolating a fungus or identifying a pest.

Chapter 10 Workbook

1. True/False—Information about a tree's history and symptoms gained from a home owner can always be considered accurate.
2. If a tree is not well suited for the site in which it has been planted, it may become ______________, predisposing it to other problems.
3. A common mistake in diagnosis is to carefully examine the aboveground portion of the tree, while ignoring the ______________.
4. True/False—If a tree declines or dies within the first year following installation, a likely cause is excess or insufficient water.
5. Leaf scorch, girdling roots, and mineral deficiencies are examples of ________________ disorders.
6. Name five causes of physical or mechanical injuries to trees.

   a.

   b.

   c.

   d.

   e.
7. Insect damage to trees is usually the result of feeding or ______ ________________.
8. Name five insect pests of trees with chewing mouthparts. Name five with piercing or sucking mouthparts.

Chewing	**Piercing/Sucking**
a.	a.
b.	b.
c.	c.
d.	d.
e.	e.

9. Insects that carry plant pathogens are said to be _______________.
10. True/False—Mites are not actually insects.
11. Microscopic roundworms that feed on trees and may carry diseases are called ______________.
12. Name the four factors required for a tree disease.

    a.

    b.

    c.

    d.
13. True/False—Vascular diseases of trees are rarely fatal.
14. True/False—Diseases that affect only the foliage of a tree may not be a serious problem unless defoliation occurs in several consecutive years.

15. True/False—Most fungi cause plant disease.
16. True/False—The pathogens that cause plant diseases are primarily fungi.
17. Fireblight is an example of a disease caused by a _______________.
18. ________________ is the chemical inhibition of growth and development of one plant by another.
19. Name four common air pollutants that can harm trees.
    a.
    b.
    c.
    d.
20. Curling and cupping of the foliage, and parallel venation, are common symptoms of_______________ damage.

MATCHING

____ witch's broom	A. swollen plant tissue, often insect or mite induced
____ vector	B. carrier of pathogens
____ canker	C. localized dead tissue, often shrunken and discolored
____ gall	D. abnormal growth of multiple secondary shoots
____ stunting	E. may predispose a plant to other problems
____ stress	F. causal agent of disease
____ pathogen	G. natural chemical inhibition of growth
____ leaf spot	H. reduced growth
____ allelopathy	I. dead spots on the foliage

CHALLENGE QUESTIONS

1. Outline the steps to be taken in the process of diagnosing a tree problem.

2. List the most common disease and insect problems in your region and categorize them by severity.

3. Explain the procedure used in collecting plant samples to be sent to a laboratory for diagnosis.

4. Why is pollution damage often difficult to control? What other disorders have similar symptoms?

5. Describe the ways that humans contribute to the reduced life span of trees in the urban environment.

SAMPLE TEST QUESTIONS

1. A condition characterized by a cluster of dwarfed shoots on affected twigs is called
   a. witch's broom
   b. anthracnose
   c. chlorosis
   d. verticillium wilt
2. Twig dieback from periodical cicadas is primarily a result of
   a. ovipositing or egg laying
   b. adults feeding on the foliage
   c. larvae feeding on the roots
   d. feeding-induced galls on the twigs and foliage
3. Plant damage associated with a sap-feeding insect pest might appear as
   a. leaves that have been skeletonized
   b. distorted leaves or shoots
   c. leaf mines or blotches
   d. webs or tents in the tree
4. Scale damage to plants is the result of
   a. fungal spore growth depleting xylem reserves
   b. phloem-feeding insects causing a loss of vigor
   c. vascular damage from fungal invasion
   d. a physiological disorder due neither to insects nor to disease
5. Damage caused by nonliving factors tend to be
   a. uniform with definite borders
   b. uniform but generally not affecting the new growth
   c. random and concentrated on the new growth
   d. random with irregular borders

Other Sources of Information

(See pages v–vi for complete bibliographic information.)
Harris et al., 1999. *Arboriculture: Integrated Management of Landscape Trees, Shrubs, and Vines.*
Johnson and Lyon, 1991. *Insects That Feed on Trees and Shrubs.*
Lloyd, 1997. *Plant Health Care for Woody Ornamentals.*
Sinclair et al., 1987. *Diseases of Trees and Shrubs.*

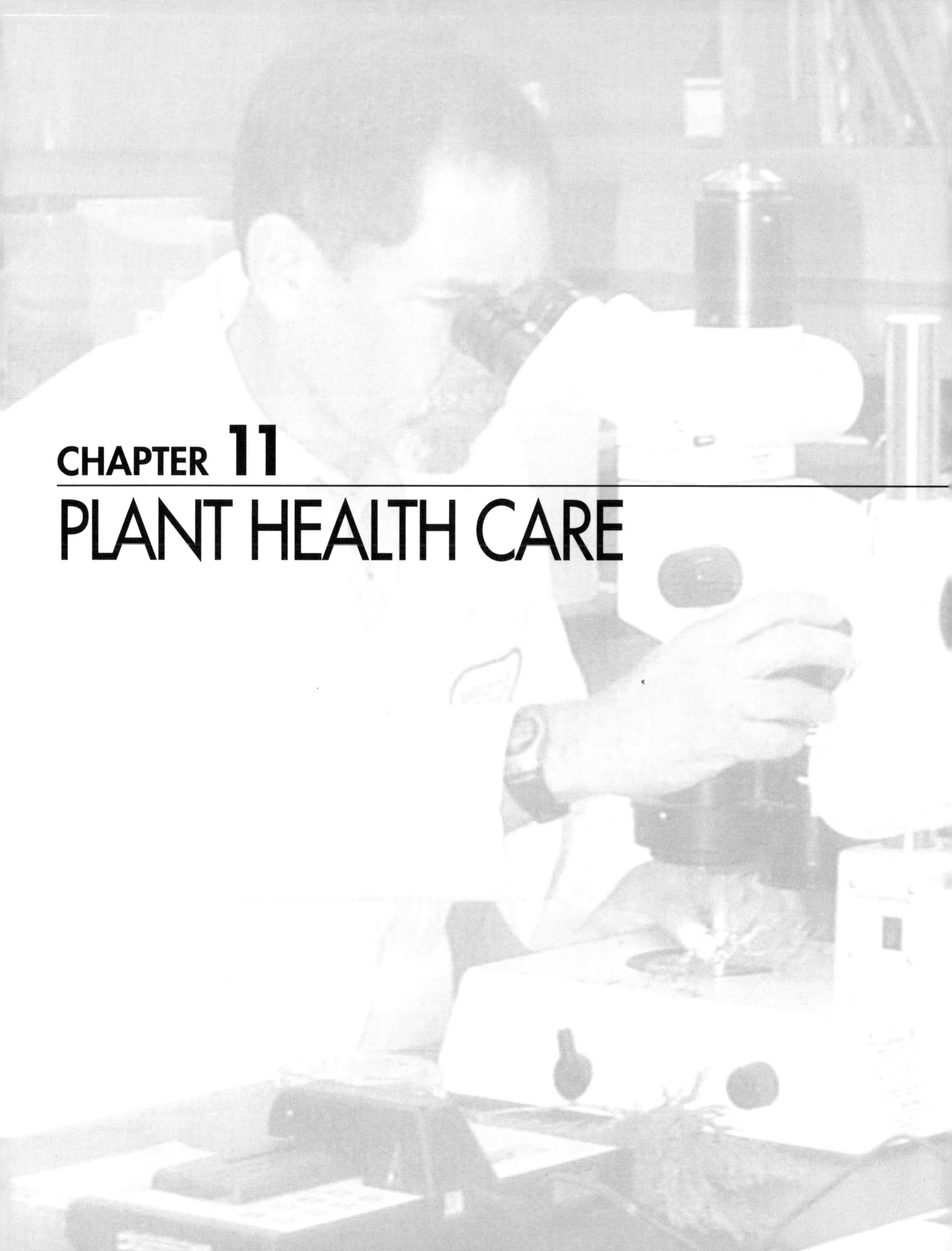

CHAPTER 11
PLANT HEALTH CARE

CHAPTER 11 PLANT HEALTH CARE

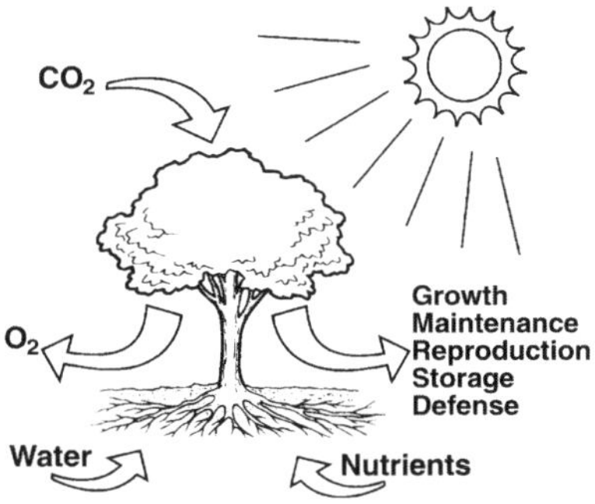

CHAPTER 11
PLANT HEALTH CARE

Objectives

1. Explain the philosophy of Plant Health Care (PHC) and describe how it differs from Integrated Pest Management (IPM).
2. Understand the concept of the appropriate response process (ARP) and how it is used in diagnosis and treatment of plant health problems.
3. Compare and contrast some different ways in which plants allocate resources depending on seasons and environmental stresses.
4. Become familiar with multiple pest control or treatment options and the advantages and limitations of each.

Key Terms

allelochemicals
appropriate response process (ARP)
botanicals
cellulose
contact insecticides
cultural controls
fungicides
horticultural oils
insect growth regulators
insecticidal soaps
insecticides
Integrated Pest Management (IPM)
lignin
microbial extracts
monitoring
mortality spiral
pest resurgence
pesticides
phenols
Plant Health Care (PHC)
resistant varieties
resource allocation
secondary pest outbreak
stress
systemic
tannins
thresholds
translocated
vigor
vitality

INTRODUCTION

There was a time when the commonly accepted approach to keeping plants healthy was a broad application of chemicals to control insect and disease problems. Today, sound plant management principles are based on maintaining plants in good health, and most tree care professionals understand that tree health problems are usually the result of several stress factors. Recommendations for broad applications of insecticides or fungicides are far less common. While pesticides play an important role in plant health management, they also have limitations. Besides environmental and health concerns that may be associated with the use of pesticides, pest control may be limited and temporary if other control measures are not integrated into the program. The options arborists have in treating problems after they develop are limited, though, and emphasis has switched to health maintenance and minimization of stress.

DEFINITION AND PHILOSOPHY

There are many definitions of **Plant Health Care (PHC)**, but the theme remains consistent. PHC is a holistic and comprehensive program to manage the health, structure, and appearance of plants in the landscape. The trees under the care of arborists are part of a larger ecosystem. Most often, they are elements of a landscape that includes shrubs, herbaceous plants, and turfgrass. The trees interact with, and sometimes compete with, other plants in the landscape. They may compete, for example, with turf for light above ground, and water and minerals below ground. In addition, the cultural practices for one component of the landscape can impact the health, growth, and development of the other plants.

Arborists have come to realize that they cannot view trees in isolation; they must consider the entire system. Consider the example of a golf course. Trees may be growing in a "landscape" of highly maintained bentgrass. Turfgrass irrigation might cause root rot problems on trees, frequent applications of high-nitrogen fertilizer could lead to elevated populations of foliage-feeding insects, and mowing procedures may damage tree trunks. On the other hand, the golf course superintendent will probably be frustrated by tree shade that limits light penetration to the turf, and perhaps limbs of some of the trees interfere with play. Every maintenance option for either trees or turf can impact the health of nearby plants. Thus, the arborist and golf course superintendent need to work together to arrive at maintenance practices that will allow trees and turf to thrive in the same environment.

In a suburban landscape, it is common for home owners to contract with a tree service for maintenance of their trees, a lawn care company to manage their lawn, and a landscape maintenance firm to care for the other elements of the landscape. Without a coordinated effort, the cultural plant care practices of each company could impact the effectiveness of the others. For example, if each company is applying fertilizer, the potential exists for excess application, which could lead to increased pest populations, unwanted growth, excess fertilizer salts in the soil, waste due to leaching, and groundwater pollution. This approach is inefficient and costly for the home owner, and perhaps the environment.

Plant Health Care (PHC) is a proactive approach that, ideally, begins with the design of the landscape and selection of plants. Many plant health problems can be avoided if plants are properly matched to their environments. In addition, if trees are planted properly and well cared for in their earlier years, their maintenance needs will be less as they mature. Unfortunately, arborists are usually called in long after these early stages, when trees are in decline. One of the main challenges of Plant Health Care is education of both green industry professionals and clients.

What Is a Healthy Plant?

Plant Health Care is the management of plant appearance, structure, and **vitality.** The appearance of landscape plants is a legitimate objective because landscapes are planted and valued as much for their aesthetic qualities as their functional utility. Structural integrity is extremely important when it comes to trees, not only for their preservation but also for safety issues. This subject is given more attention in Chapter 12 of this guide. But what constitutes a healthy plant? Many plant care professionals differentiate between vitality and **vigor**. Vitality is sometimes defined as a plant's ability to deal effectively with stress. Vigor deals with the plant's inherent genetic capacity to resist stress. A plant's adaptability to various soil types, moisture conditions, cold or heat, and other factors is what defines the plant's vigor. The only way to take advantage of vigor is to select plants well matched to the environment and resistant to pests. Most plant maintenance professionals are engaged in promoting the vitality of plants because they are usually called in long after selection and planting have taken place.

Some arborists and other landscape professionals, researchers, and consumers equate rapid growth with a healthy plant. The success of various treatments and practices has usually been measured by increased growth in height, trunk diameter, root mass, or total leaf area (Figure 11.1). As researchers have gained a better understanding of plant physiology and **resource allocation**, it is apparent that other aspects of plant vitality are often more important.

Trees and other plants produce sugars, the basic energy building blocks, through the process of photosynthesis. This energy is allocated between four primary functions: maintenance (including reproduction), growth, storage, and defense. The allocation to each function varies by season and environmental conditions, and even by genetic strategies. For example, some trees grow rapidly, putting a relatively small percentage of resources toward defense. Others grow at a slower rate, but are strong compartmentalizers (Figure 11.2).

Photosynthesis requires light, water, and a few essential

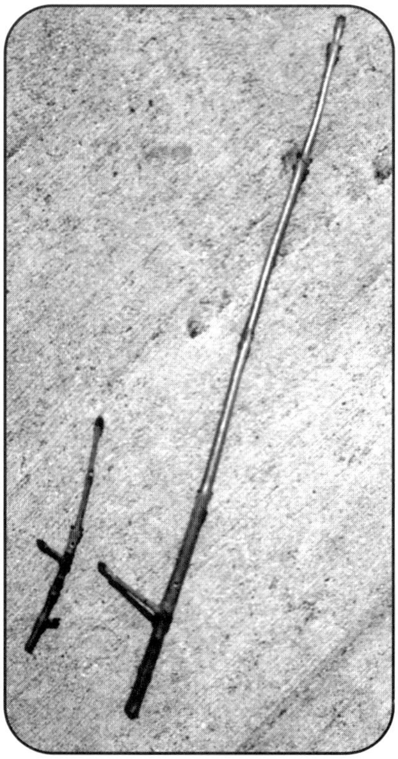

Figure 11.1 Growth rate is just one indication of tree vitality. The twig on the right shows much greater twig extension than the twig from the stressed tree on the left.

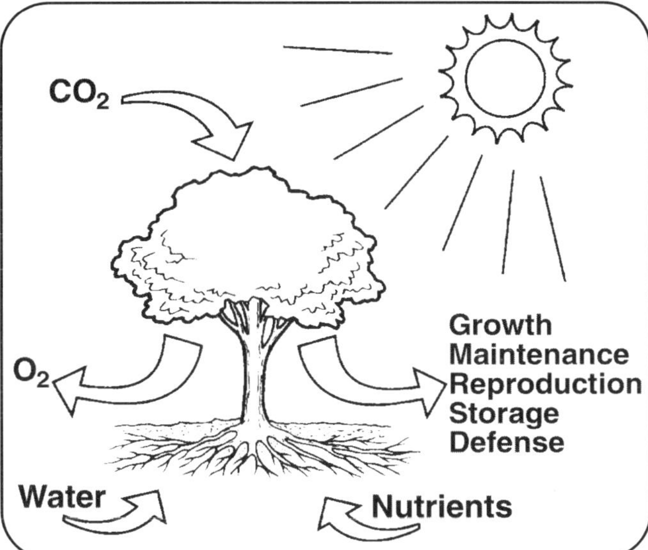

Figure 11.2 Resource allocation. Trees carefully budget their limited supply of resources among the various processes that must be supported, including growth, maintenance, reproduction, storage, and defense. Not all processes can be supported fully at the same time.

minerals. If any of these fundamental ingredients are limited, the plant will be stressed. Any factor that limits a plant's ability to acquire these resources or leads to excessive amounts of these resources can be considered a **stress** factor. Stress factors are often directly related to soil and other environmental conditions (for example, pH, drought, or excess water). Some stress factors reduce a tree's ability to photosynthesize, decreasing the amount of sugar produced. This can cause the tree to go into a survival mode, redirecting resources away from growth, storage, and defense, and toward maintenance (Figure 11.3). On the other hand, mild stresses can redirect resource allocation with the effects of actually increasing drought or pest resistance.

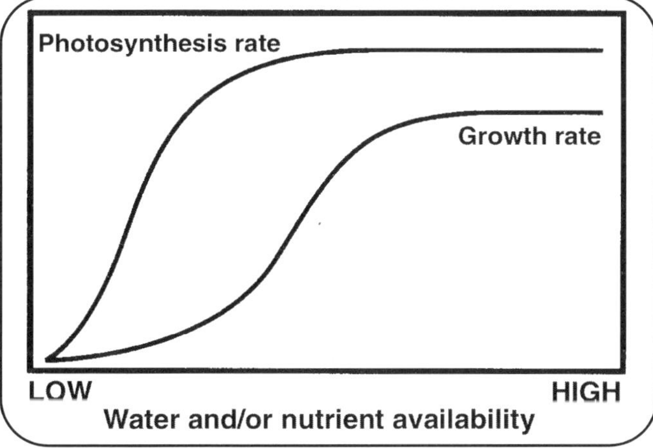

Figure 11.3 Effect of stress on growth and photosynthesis. The growth of trees is quite sensitive to the availability of moisture and nutrients, and becomes limited by even mild deficiencies. Photosynthesis is much more resilient and does not become limited until nutrient and moisture deficits become more severe.

Plant Defense Mechanisms

All trees possess certain traits that deter some insects and diseases. Some have thorns, spikes, or tiny hairs on the leaves that are a physical deterrent. Some have a thick, toughened cuticle. **Cellulose** and **lignin** in tree cells cannot be digested by many insects, animals, and even some pathogens. Some of the same processes involved in compartmentalization of decay can also resist insect damage and disease spread.

Trees also produce **allelochemicals** such as **tannins** and **phenols** that have toxic or deterrent effects on certain insects. In fact, we use extracts of some of these allelochemicals for pesticides. Nicotine, pyrethrin, and rotenone are examples of these natural insecticides. These allelochemicals may be the reason most insects are host specific. Insects have developed special enzymes that allow them to detoxify specific compounds.

Recent studies have investigated the relationship between cultural influences and allelochemical production in trees. For example, moderate drought stress can increase levels of allelochemicals, perhaps boosting the defenses of the tree. Conversely, sun-adapted trees grown in shade have reduced levels of allelochemicals and lowered resistance to some pests. Much research has concentrated on the effects of nitrogen fertilization on host–pest relationships. Researchers have observed that rapidly growing trees are sometimes less resistant to certain insects and diseases. This finding is contrary to the traditional belief that a rapidly growing tree is a healthier tree. One explanation is that photosynthate is diverted from defense (production of allelochemicals) to increased growth of succulent tissues. Increased growth without a corresponding increase in photosynthesis may stress the plant.

Trees can adjust to moderate levels of stress in a variety of ways. When available water is limited, stomata may close, the levels of dissolved substances in cells may change to maintain turgor pressure, and the root:shoot ratio may increase. Longer-term moisture stress triggers more dramatic responses such as leaf drop. Trees respond to low light levels by increasing shoot growth toward the light source and by adjusting leaf size and thickness to maximize photosynthesis. If nitrogen is deficient, trees may decrease shoot growth and increase root growth to maximize mineral uptake and availability for existing foliage. They may also produce smaller, thicker leaves. This has the effect of maintaining high rates of photosynthesis with decreased growth (Figure 11.4).

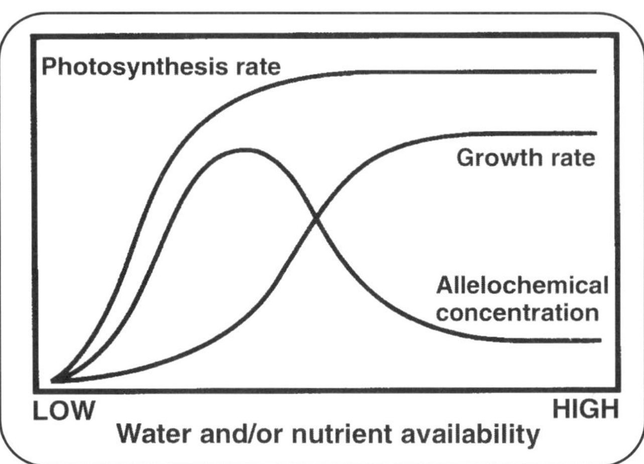

Figure 11.4 Moderate nutrient and moisture stress does not impact photosynthesis but does limit growth, making carbohydrates available to support other processes, such as defense.

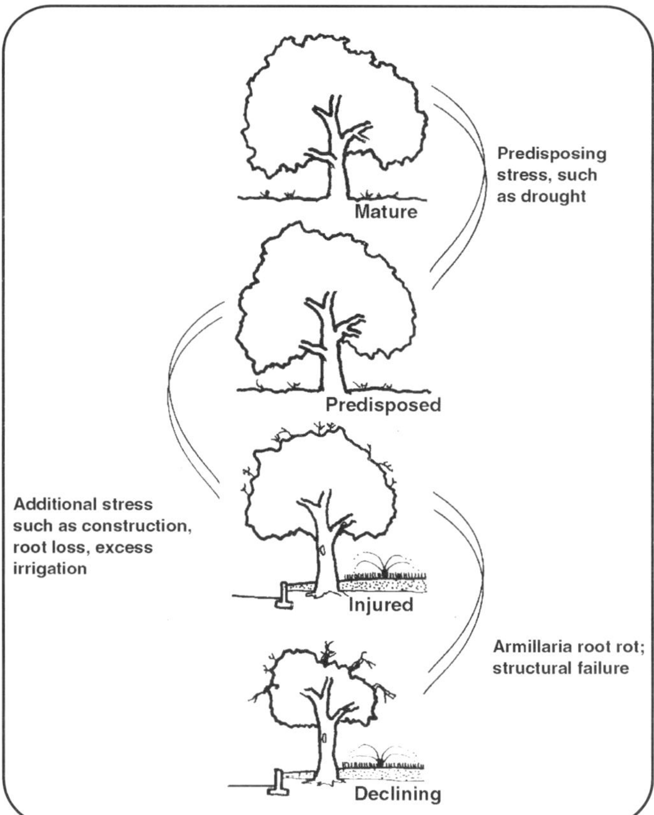

Figure 11.5 The mortality spiral illustrates how stress factors compound, predisposing a tree to additional problems and leading to decline.

While mild levels of stress can sometimes increase resistance to certain insect pests, stress has a cumulative effect, and multiple stresses can compound, putting a tree into a state of decline. Stressed trees are more likely to succumb to drought, defoliation, borers, bark beetles, or vascular wilt diseases. When a tree has entered this state of decline, it is said to be in a **mortality spiral** (Figure 11.5). As already stated, the options an arborist has for reversing the effects of stress are often limited, so the emphasis must logically be placed on avoiding or minimizing stress.

The Process of Plant Health Care

The key to the success of any PHC program is **monitoring** the plants and growing conditions. Monitoring is an important process of observing, identifying, recording, and analyzing what happens with plants in the landscape (Figures 11.6 and 11.7). Early detection and correction of situations that might cause plant stress can head off more serious problems in the future. If pest problems or other disorders are discovered, the severity and potential impact of the situation must be assessed before deciding on an appropriate response (Figure 11.8). The process of gathering information, assessing the severity and implications of the problem, determining client expectations, and deciding upon a course of action is called the **appropriate response process (ARP)** (Figure 11.9).

Figure 11.6 An important part of a successful PHC program is monitoring the plants and growing conditions. Early detection and correction of stress factors can head off more serious problems in the future.

Arborists have borrowed the concept of thresholds from **Integrated Pest Management (IPM)** applications with agronomic crops. Where plants are grown as crops, the threshold for treatment of a pest population is determined by doing a cost–benefit

Figure 11.7 Soil testing provides important information about growing conditions.

Figure 11.8 Sometimes it may be necessary to seek laboratory confirmation of findings.

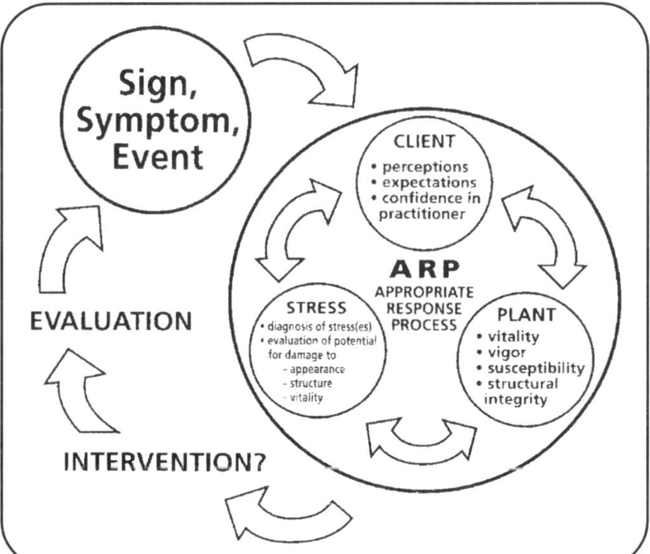

Figure 11.9 The appropriate response process (ARP) is one in which the arborist considers the tree health, stress factors, and client expectations in determining an appropriate response.

analysis of crop loss versus treatment expense. The thresholds for intervention measures on ornamental plants are less quantitative and depend on the objectives of the client. As an example, with spiders in the bathroom, the threshold population that requires immediate intervention is one. Even though spiders are considered beneficial, there is usually no tolerance for their existence in the bathroom. Likewise, although some pest or disease problems have little effect on the overall vitality of a plant, the aesthetic qualities of the plant may be temporarily spoiled, or the pest may create a nuisance. Treatment may be indicated.

Another important aspect of PHC is client education. In the example of spiders in the bathroom, treatment was called for even though the problem was primarily emotional. In some cases, it is preferable *not* to treat cosmetic problems, but rather to help the client understand the nature of the pest and the minimal impact on the tree's health. It is up to professionals to make educated decisions and avoid short-term treatments at the expense of long-term solutions. This is part of the theory of appropriate response.

TREATMENT OPTIONS

One of the lessons learned through experience with previous plant pest problems is that broad insecticide applications can reduce the populations of beneficial insects as well as pests. The result could be target **pest resurgence** of more chemical-resistant pests, in the absence of natural parasites and predators, or a **secondary pest outbreak**. Usually there is a natural, dynamic balance in the ecosystem. In fact, it is not practical to attempt total eradication of any one pest because this could lead to the demise of its predators or parasites. Then any reintroduction of the pest could lead to a repeated occurrence of the pest's attack.

Integrated Pest Management (IPM) is a systematic approach to insect and disease management. It incorporates a combination of techniques including resistant plants; cultural practices; and mechanical, biological, and chemical controls of plant pest problems. The goal is to maintain tree health while minimizing the adverse ecological impact of the intervention practices.

Resistant Varieties and Cultural Controls

The best way to deal with plant health problems is to avoid them. When possible, arborists should select resistant varieties (trees resistant to or tolerant of known insects or diseases) and avoid planting trees in sites for which they are ill suited. Subjecting a tree to stress predisposes it to other health problems. Taking this rationale one step further, if an arborist encounters a tree problem that requires extensive intervention on a frequent basis, the best alternative might be to recommend removal and replacement with a more appropriate plant for the site. An example might be a crabapple (*Malus* spp.) that annually suffers defoliation from apple scab. Most crabapples are planted for their ornamental qualities, and one that goes most of the summer with a few scabby, chlorotic leaves gracing its branches does not live up to our aesthetic expectations. Rather than annually spraying a series of fungicidal protectant sprays at considerable expense, it might make sense for the client to consider replanting with a cultivar resistant to apple scab. Of course, the expectations and feelings of the client will be a key factor in that decision.

Cultural practices are the essence of Plant Health Care and really are simply a matter of good horticulture. If a tree is subject to fungal problems on its foliage, it is wise to avoid irrigation that constantly wets the leaves. If two species are alternate hosts to a specific disease, they shouldn't be planted in proximity to one another. Other **cultural controls**, such as good sanitation, mulching, and providing adequate water, can help prevent or minimize plant health problems.

Chemical Controls

Chemicals have been used to control insect and disease problems on plants for centuries, with varying degrees of success. In the last few decades, a great deal of research has been done in developing **pesticides** that are more target-specific and less toxic to other species or the environment.

The primary pesticides used by arborists are **fungicides** and **insecticides**. Although there are some exceptions, fungicides are usually applied as protectants to prevent infections from becoming established in or on a plant. Most insecticides are **contact insecticides**, killing insects that make direct contact with the spray or its residue. Some insecticides are **systemic,** that is, they are taken up by the plant and **translocated** throughout the branches and into the leaves. Systemic insecticides can provide broad coverage and can reach insects that feed in or on the foliage.

Chemical control of insects and diseases is still an important part of maintaining tree health. However, if chemicals are to be used, they must be used wisely. Pesticides should be selected to minimize their effect on nontarget organisms. Target-specific spraying is preferred to broadcast spraying. All applications must be done according to all safety regulations, according to label instructions, and in accordance with all state and federal regulations. In the United States, as in many countries, commercial pesticide application requires a license. Licensing is administered by state departments of agriculture.

Alternative Pesticides

Another pest control option is the use of an alternative pesticide. Examples include **insecticidal soaps, horticultural oils, botanicals, insect growth regulators,** and **microbial extracts.** An advantage of many of these choices is that they tend to be somewhat pest specific and less damaging to natural enemies.

Insecticidal soaps are highly refined soaps that disrupt the cell membranes of soft-bodied insects and mites. They are effective on scale crawlers, aphids, mealybugs, and spider mites. They have no residual effects. Horticultural oils have insecticidal properties because they suffocate insects or disrupt their membranes. Horticultural oil applications are divided into two categories: summer oil applications and dormant oil applications, which vary based on timing and rates. Dormant oil sprays are applied before leaf break and are highly effective on some pests without damaging beneficial insects. Summer oils can be sprayed on the foliage of some plants (although other plants may suffer phytotoxicity injury).

Botanical pesticides are plant extracts used for insecticidal purposes. Botanicals range from highly toxic to quite mild, for both insects and people. Insect growth regulators are synthetic compounds that act like insect hormones, disrupting the molting or growth processes. They have several uses, including baits for trapping and monitoring insect populations. They have also been used to control populations by disrupting the molting process of certain insects, or by making the adults sterile.

Microbial pesticides contain insect pathogens or their products that are derived from extracts of bacterial pathogens of insects. The most common example is *Bacillus thuringiensis (Bt)*, a bacterium deadly to certain insects. Different strains are effective against different groups of insects. One type is used for leaf-feeding beetles, another affects caterpillars, and a third is effective against mosquito larvae. These products have shown almost no toxicity to nontarget organisms or the environment (Figure 11.10).

Biological Controls

As mentioned previously, many insects live in a natural balance with predators and parasites. One objective

Figure 11.10 Biological controls are an important alternative in insect pest management. The Japanese beetle grub on the left is normal; the one on the right is infected with "milky spore."

of pest control should be to conserve or augment populations of beneficial insects. In some cases where the pest population has far exceeded natural enemy populations, predators or parasites have been artificially reintroduced to establish the balance. Some natural predators and parasites are available commercially. In most cases, however, more knowledge of pest and predator life cycles is needed to make this strategy effective (Figures 11.11 through 11.13).

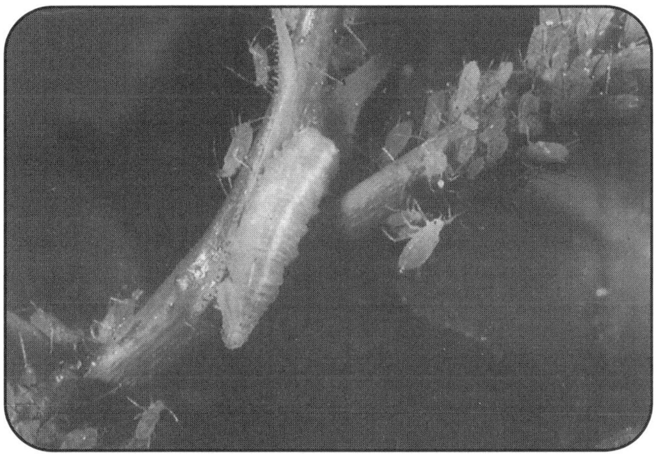

Figure 11.11 Natural predation helps keep pest populations in check, as with this syrphid larva eating aphids on a rose plant.

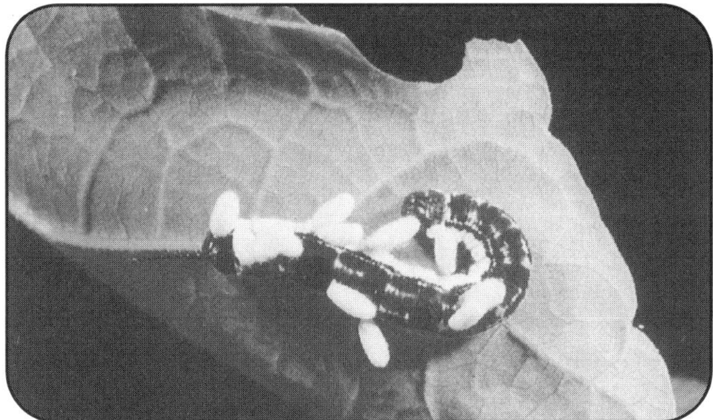

Figure 11.12 Catalpa sphinx with parasites, an example of natural pest population control.

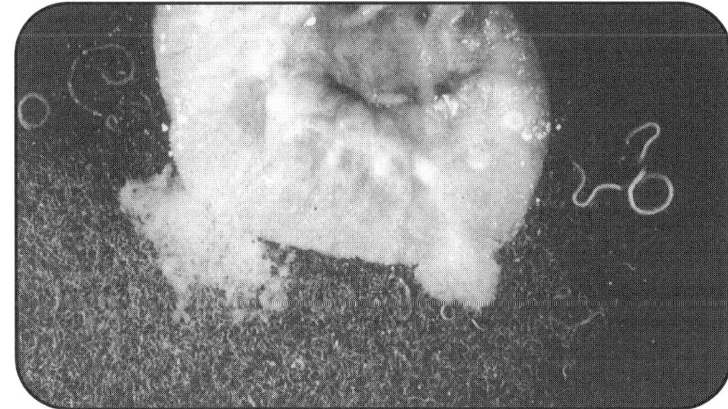

Figure 11.13 This Japanese beetle grub has been parasitized by nematodes (seen here as fine white lines.)

Mechanical Management

On a small scale, mechanical management can sometimes be the most practical alternative for pest control. Picking bagworms off a single evergreen shrub or small tree, for example, might take only a few minutes. Sometimes, mechanical management could consist of installing physical barriers to keep pests away. Another example is pruning out infested plant parts, such as might be done with fall webworm or fire blight.

Chapter 11 Workbook

1. True/ False—Plant Health Care and Integrated Pest Management are essentially the same thing.
2. __________ is sometimes defined as a plant's ability to grow in its particular environment.
3. A plant's genetic adaptability to various soil types, moisture conditions, cold or heat, and other factors is described by the term ________.
4. Plant resources, produced through photosynthesis, are allocated between five primary functions:

   a.

   b.

   c.

   d.

   e.
5. ___________ can be described as any factor that limits a plant's ability to acquire sufficient light, water, or essential minerals.
6. The ___________ and ___________ in tree cells are indigestible to many insects and other animals, and even some pathogens.
7. Trees also produce _________________ such as tannins and phenols that have toxic or deterrent effects on certain insects.
8. One of the reasons that most insect pests are somewhat host specific is that insects have developed special ___________ that allow them to detoxify specific compounds in certain plants.
9. True/False—Researchers have observed that rapidly growing trees are sometimes less resistant to certain insects and diseases.
10. _____________ is the process of observing, identifying, recording, and analyzing what happens with plants in the landscape.
11. The process of gathering information, assessing the severity and implications of the problem, determining client expectations, and deciding upon a course of action is called the _________________ __________ __________.
12. _______________ ______ _____________ is a systematic approach to insect and disease management that incorporates a combination of techniques including resistant plants, cultural controls, mechanical management, and chemical controls of plant pest problems. It is considered by many to be a part of Plant Health Care.
13. When possible, arborists should select trees that are _______________ to known insects or diseases.
14. Insecticides that are __________ are taken up by the plant and translocated throughout the branches and into the leaves.
15. True/False—Insecticidal soaps disrupt the cell membranes of soft-bodied insects and are effective on some scales, aphids, mealybugs, and spider mites.
16. True/False—Horticultural oil applications are always safe to use on trees in leaf because they have no phytotoxic properties.
17. _________ _______ _____________ are synthetic compounds that act like insect hormones.
18. _______________ _________ are derived from certain bacterial pathogens of insects.

19. True/False—*Bacillus thuringiensis (Bt)* is a bacterium that is deadly to certain insects.
20. Many insect pests live in a natural, dynamic balance with ____________ and _____________ that control pest populations.

CHALLENGE QUESTIONS

1. Select two species of trees in your area that have different genetic strategies for resource allocation. Compare and contrast the advantages and limitations of each.
2. Describe some of the defense mechanisms found in trees.
3. Explain how different treatment and control strategies can be implemented to control pest problems on plants.

SAMPLE TEST QUESTIONS

1. Plant health care is a comprehensive program to manage
   a. insects and diseases of plants
   b. tree health without the use of pesticides
   c. the appearance, structure, and vitality of plants
   d. pests, pathogens, and abiotic disorders of trees
2. The mortality spiral describes the
   a. process of infection and spread of disease in a tree
   b. cumulative effects of stress causing decline of a plant
   c. process in which pesticides eradicate both pests and beneficial insects
   d. allocation of resources between growth, storage, and defense
3. The process of gathering information, assessing the severity and implications of the problem, determining client expectations, and deciding upon a course of action is called
   a. the appropriate response process
   b. Integrated Pest Management
   c. the cultural control mechanism
   d. Plant Health Care
4. A systemic insecticide is one that
   a. kills all living organisms
   b. kills insects on direct contact
   c. is translocated throughout the plant
   d. has no harmful effect on the environment

5. Releasing predators or parasites of an insect pest is an example of a
   a. cultural control
   b. mechanical control
   c. biological control
   d. chemical control

Other Sources of Information

(See pages v–vi for complete bibliographic information.)

Lloyd, 1997. *Plant Health Care for Woody Ornamentals.*

CHAPTER 12

TREE ASSESSMENT AND RISK MANAGEMENT

CHAPTER 12 TREE ASSESSMENT AND RISK MANAGEMENT

CHAPTER 12
TREE ASSESSMENT AND RISK MANAGEMENT

Objectives

1. Discuss the process of tree risk assessment, the factors to be considered, and some of the defects that can create potential hazards.
2. Explain why decay can be a significant risk factor, and discuss some of the signs of decay in trees.
3. Discuss the liability associated with tree risk management and the role of the arborist in assessing the condition of trees and making recommendations for tree care.

Key Terms

branch union	cracks	included bark	reaction wood	splits
cavity	decay	liability	risk assessment	structural defects
codominant stems	fruiting bodies	mitigation	risk management	taper
conks	hazard potential	negligence	scaffold branches	target

INTRODUCTION

Evaluation of the **hazard potential** of urban trees is an important role of the professional arborist. Trees are hazardous when failure of their parts results in property damage or personal injury. Identification and correction of **structural defects** may reduce the failure potential (and, therefore, reduce risk to property or injury to people) and may also prolong the life of the tree. **Risk assessment** includes three components: the potential of the tree to fail, the environment that may contribute to failure, and the potential **target**. Each factor has an impact on the hazard rating of the tree.

In evaluating the potential of the tree to fail, the arborist must consider the species, growth habits, branch attachments, defects, condition of the root system, lean, and the history of the tree. It is also important to consider the size of the potential failure. Obviously, the damage potential is greater for large limbs than small branches.

The environment also plays a part in the potential for tree failure. Most tree failures occur during, or as a result of, storms. Exposure to winds, snow and ice loading, lightning, and rainfall should be considered when assessing trees. Other environmental factors such as soil conditions and slope can affect the likelihood of failure and should be taken into account. Determining the history of the site may also reveal some factors (for example, construction, grade changes, and trenching) that can affect the health and/or structural stability of trees on the site.

By definition, if there is no potential target, a tree does not pose a hazard. Targets may include structures, vehicles, or people. Structures are relatively easy to assess because they are fixed. However, in determining the likelihood of a failed tree or branch striking a person, one must consider the frequency and intensity of use of the site. A tree that poses unreasonable risk on a school playground has greater potential to cause injury than a defective tree in an open field.

Evaluating trees for their potential to fail is a component of **risk management**. Almost all trees pose a risk of failure. As trees grow larger and more mature, the degree of risk increases. A challenge for arborists is to manage trees at a level of risk acceptable to the client. That level of risk will have to be determined by the property owner or manager. Arborists frequently must make recommendations about the level of risk trees represent, the potential to reduce that risk, and the advisability of removing trees. Decisions must be based on a thorough assessment using the best information available. There are times, however, when it is advisable to consult a specialist in plant pathology or hazard tree assessment.

IDENTIFYING POTENTIAL HAZARDS

The ability to predict tree failure is limited, but with proper training arborists can learn to identify characteristics that often result in failure. Defects are not always visible, especially those inside the tree or beneath the ground, and the forces of nature are unpredictable. With experience, however, one can learn to recognize patterns of failure that will help in recognizing risk factors. For example, forest edge trees are typically more prone to failure than those deeper in the woods, surrounded by other trees. Characteristics such as bowed trunks and asymmetric crowns may be indications of structural problems.

General Inspection

Inspecting trees for hazards involves a great deal more than a casual stroll among them. Tree inspection must be a systematic process. Diagnosing tree hazards requires a fundamental knowledge of tree structure and physiology. It takes a trained eye to discern the difference between a minor flaw and a possible hazard.

When performing an inspection, it is important to stick to a systematic and constant process. This will help ensure that you inspect every portion of the tree and do not miss anything. First, assess the tree as a whole. Look for dieback in the crown. Take note of any lean or branches that extend beyond the rest of the crown. Then inspect the trunk, the root crown, and the root zone. The root collar is a critical region for determining sound structure and should be given close attention. Failure at this point is common and can be catastrophic. Look for a visible root flare, with solid, well-developed roots without decay (Figure 12.1). Finally, examine the canopy of the tree. By sticking to a system, you are more likely to perform a more thorough inspection.

Structure

- Examine the branch angles and branch attachment of the major **scaffold branches**. Branches should be smaller in diameter than their parent branches. If they are equal in size or larger than the parent branch, they are much more likely to fail. Check for proper branch bark ridge formation. This is an indication of a well-attached branch.

Figure 12.1 Systematically inspect trees for flaws or defects that could be risk factors.

- Watch out for **codominant stems**. These unions have a high risk of failure. Codominant stems often have **included bark** within the **branch union**, making them structurally unstable (Figure 12.2).
- Stems with good **taper** (stems that are significantly larger at the base and smaller toward the end) tend to be stronger than those without good taper. Trees that have grown in dense shade such as a forest tend to have long, straight stems

Figure 12.2 This tree failure occurred in the branch union between codominant stems.

with little taper. If the protection of the surrounding trees is removed, they are more likely to fail.

- Many trees have a natural lean. Trees often produce **reaction wood**, providing more stability to compensate for the lean. Some trees, however, may lean due to recent upheaval, and could be unsafe. Also, a lean combined with other defects may be cause for concern.
- Look for signs of vigorous growth. Some tree species are better compartmentalizers of decay than others are. Growth rate and good compartmentalization are important in the reduction of hazards. A tree that closes and compartmentalizes a wound rapidly may be less likely to become a hazard than one that does not. Although compartmentalization is important, it does not guarantee structural stability. Extensive decay, compartmentalized or not, is a risk factor (Figure 12.3).

Figure 12.3 This tree has multiple wounds and is showing signs of closure and compartmentalization. Extensive decay, however, is a serious risk factor.

- Keep in mind that tree health and structure can be mutually exclusive. Green, full-crowned trees can fail due to structural defects.

Potential Problems (Faults and Defects)

- Dead branches within the canopy of a tree are probably the most obvious potential hazards. The risk of damage or injury depends on the size of the dead branch and the location of the tree.
- Watch for longitudinal **cracks** or **splits** on the trunk or major branches. Cracks that start at a branch union may be especially hazardous. Cracks on decayed stems or branches may indicate imminent failure (Figure 12.4).

Figure 12.4 Longitudinal splits or cracks can pose a hazard. At times, pruning and cabling and/or bracing can mitigate the risk but might not eliminate it.

- An observant arborist may be able to detect internal defects where the stem is out of round. Asymmetric shapes may be caused by the tree's formation of reaction wood and may be an indication of an internal problem.
- Codominant stems, especially with included bark, can often be failure points. Multiple branch attachments at one point on a stem can also be a defect.
- Branches or stems that lack taper, especially if weight is concentrated near the end, are prone to failure.
- Externally visible defects include cankers, galls, and wounds. Each could be minor or the start of a significant problem. If such defects are discovered, further investigation is warranted.
- Cracks or separations in the soil may indicate soil heaving from excessive movement of the roots, which can be a warning sign for failure, especially if the tree is leaning.
- Some defects are humanmade. Wounds, weak or damaged limbs, and decay may be the result of poor pruning or other misguided practices. Root loss, not always obvious to the untrained observer, can result from construction, trenching, grade changes, and soil compaction (Figure 12.5). Any of these activities can lead to a decline in health or structural stability that may require action.

Figure 12.5 Wounds caused by poor maintenance practices, such as lawn mower damage, can worsen with decay. This tree also shows no visible root flare, which may indicate root loss.

Although this list is not comprehensive, it does note some of the most common tree defects that can lead to failure. Experienced arborists learn to recognize these and other potential problems when assessing trees.

Decay and Discoloration

Decay is perhaps the most insidious defect within a tree. Decay can cause significant strength loss. The larger the area of decay, the thinner the remaining sound wood to support the tree, and the greater the strength loss. There are a number of formulas to estimate the amount of a trunk that is decayed. Most formulas are based on determining the proportion of sound wood relative to stem/branch diameter. While the formulas vary, most experts agree that a threshold of 30 to 35 percent loss requires that some action be taken. If there are large **cavity** openings or other aggravating factors, the threshold drops to 20 to 25 percent. It is difficult to determine how much sound wood is required to support a given tree. Variables include the strength of the wood, the canopy size and configuration, defects, and any number of environmental factors. Experts agree, however, that the extent of decay is a critical component.

The problem is that decay is not always obvious upon external inspection of the tree. The tree may appear to be solid, structurally sound, and may have a thick, green canopy, yet can have significant decay inside. It is important to recognize signs of internal decay.

- Look for open wounds or cavities. They may be the starting points of decay. Also, dead branches may contain decay, which can spread into the tree.
- Cracked or loosened bark should be examined closely for signs of decay underneath.
- The presence of certain insects can indicate decay. Carpenter ants are usually associated with soft, decayed wood within the tree. Wood-boring insects can be another indicator. Look for frass, a sign of insect activity, accumulating at the base of the tree or in crotches.
- One of the best signs of decay is the presence of the **fruiting bodies** of the decay organisms. Mushrooms at the base of the tree or along major roots may indicate internal decay (Figure 12.6). **Conks** on the trunk or major branches are a sign of the decay fungi inside (Figure 12.7). Not all decay organism fruiting bodies are persistent, however. Some appear annually and disappear later.

Figure 12.6 Mushrooms at the base of the tree can indicate significant root decay.

Figure 12.7 Conks, or fruiting bodies, on a tree are a sign of decay fungi inside.

- Birds and bees can also be a sign of decay. Some birds nest in tree hollows, and some feed on the insects that may be present in the wood. Honeybees often make their hives in tree cavities. Small cavities are not a problem;

they serve as habitat for birds and small animals. Cavities that encompass most of the cross section of the trunk or branch can be a significant defect, however.

If decay is found or suspected in a tree, the next logical step is to evaluate its significance to tree failure. In some instances, the location and extent of decay is visible. More often, a little detective work is required. If decay is in the trunk of the tree, it can sometimes be detected with a rubber mallet or a very small-bore drill. These simple tools, however, can provide only limited information, and they are not very accurate for determining the extent of the decay. Tree inspections normally involve a visual assessment from the ground, but there may be situations that warrant further inspection.

- If the presence of decay may create a significant risk of failure, the extent of the decay may need to be determined. A number of decay detection devices are available to arborists. Most are effective and reasonably accurate when used properly, but some are somewhat invasive. Each requires a base level of training to properly gather data and interpret the results. Reliable assessment requires expertise in the technology of the instruments and methods used.
- If decay is suspected at the root flare or in the major support roots, a root collar excavation may be necessary to ascertain the extent. Root excavations are delicate and should be performed only by a qualified person. If there is any doubt when assessing decay within a tree, it is best to call in a consulting arborist who specializes in risk assessments.
- Defects in the upper crown may need a closer look via an aerial inspection. A climber or worker in a lift device can probe cavities and look for cracks and defects not visible from the ground.

RISK ASSESSMENT AND MANAGEMENT

Competent assessment of tree risk requires not only a strong foundation in tree biology but also a fundamental comprehension of the structure–function relations of trees. Without this, potential problems may go unrecognized or inadequately evaluated.

A strategy should be developed for the assessment process. All observations, measurements, and recommendations must be documented. It is best to use a hazard assessment form for consistency. The advantage of using a standard form is that it reduces the likelihood of missing any tree or site factors that could affect the evaluation. The assessment form should identify the tree genus and species as well as the common name. The size, location, and a general description should be recorded, along with the surroundings and a description of any potential "targets" that could be affected in the event of tree failure. It is important to consider the environment of the tree, including prevailing winds, microclimates, surrounding trees, and other plants (Figure 12.8). The history of the tree, including maintenance and management practices of the tree and its surroundings, should all be documented, if known.

Figure 12.8 It is common for entire trees to uproot, especially in heavily irrigated sites. Environmental conditions must be considered in risk assessments.

The tree should be systematically assessed, starting with the tree as a whole, then taking a closer look at its components. In many cases, it will be necessary to investigate further than what can be accomplished visually from the ground. It may be necessary to inspect from an aerial lift or have a climber ascend the tree for a closer look. Some trees may require a more in-depth evaluation of the root collar.

There are a number of rating systems for evaluating the hazard potential of trees. Matheny and Clark established one of the best known in *A Photographic Guide to the Evaluation of Hazard Trees in Urban Areas.* This system uses a numeric formula to quantify the risk of each tree. (The book includes a sample, standardized form for risk assessment.) The Matheny and Clark system, or other rating systems, can be integrated into a risk management program by prioritizing the trees by their need for attention, hazard potential, and exposure.

The final decision in risk management usually does not lie with the arborist. The property owner or manager must decide what level of risk is acceptable. The arborist's role is to convey the potential for failure. If a client's goal is principally safety (for example, a defective tree is overhanging the house), then tree removal or other risk-mitigating procedures might be in order. If tree preservation is a high priority and tree failure is not imminent, the client may choose to accept some risk (Figure 12.9).

Figure 12.9 All trees pose a risk, some more than others. Risk management entails the property owner or manager deciding what level of risk is acceptable. The arborist can assess risk and make recommendations.

MITIGATION OPTIONS

Part of risk assessment is evaluating the hazards and making recommendations for abatement. Some trees will pose an unacceptable risk of overall failure that cannot be abated except by removal of the tree. On a property or management area with many mature trees, though, even the immediacy of removal may have to be prioritized. If a tree has been condemned due to structural hazards, no competent manager will want to delay its removal—the liability risk is too great. On the other hand, if dozens of trees have been slated for immediate removal, a priority system must be put in place to proceed first with those that received the greatest hazard rating. A risk-rating system is helpful for the client to understand the potential for failure.

Sometimes there are management options short of tree removal to abate the hazards. **Mitigation** is the process of reducing the risk potential. Dead or broken branches can be removed from a tree's crown. The primary abatement treatment is pruning to either remove defective parts or reduce weight on them. Although risk of failure cannot be eliminated, often it can be reduced to an acceptable level. Pruning can reduce branch end weight, decreasing the likelihood of failure. If a tree has a split or codominant stems, removal of one side may be an option. Other options might include removing the target or fencing off the area and posting signs.

It may be advisable to install cables or braces to support weak branches, perhaps in conjunction with pruning (Figure 12.10). Such branches and any hardware installations should be checked regularly. If a tree has significant value, a lightning protection system could be installed for additional protection against lightning damage.

Figure 12.10 Installation of cables for added structural support is one mitigation option.

Treatment options for cavities or hollows are limited. Past practices involved removing the decayed wood and filling the cavity with various rigid fillers. More current research indicates that filling cavities may do more harm than good. Further, decay usually develops in the interface between filler and tree. In addition, the fill material may not strengthen and support the tree as much as the new callus growth that develops around the wound. If a tree has good vitality, it may maintain structural integrity by production of new wood that forms after injury. Removal of decayed wood from a cavity has little effect. However, if healthy wood tissues are damaged in the attempt to remove the decay, the tree's ability to control the spread of decay may be reduced. In most cases, it is better to leave the cavity alone. If the tree is not growing vigorously and the cavity is severe, the tree may need to be removed or reduced significantly in size.

Another option for trees of lower risk is continued monitoring. With good management practices such as proper irrigation, proper fertilization, aeration, and mulching, a tree may respond favorably and improve in condition. An example would be good callus and wood production following wounding. Trees that have been pruned or had cables and/or braces installed should also be monitored and reassessed in the future. Weight development on weak areas should be monitored and managed as appropriate, usually by judicious pruning.

Risk evaluation is usually not a one-time occurrence unless the trees in question have been removed. Whenever trees are being managed on a property, in a park system, or within a municipality, they should be put on an evaluation cycle that takes into account their size, age, overall condition, and the management practices. The frequency of evaluation depends on budget and the level of risk acceptable to property owners or administration. Most risk management programs call for annual inspections.

LIABILITY AND NEGLIGENCE

When a person has suffered injury or damages, liability must be identified and placed. Testing the circumstances for negligence is often the first step. **Negligence** is the failure to exercise due care. Negligence occurs when somebody fails to perform a duty or obligation recognized by law for the protection of others against unreasonable risks. An example of negligence might be if a tree trimmer cuts branches from a tree, and the branches fall and injure a person below. In this situation, the tree trimmer could be found negligent because he or she did not take proper precautions to protect individuals or keep them out of the vicinity. Other parties, such as the employer or property owner, could also share some liability.

Courts must determine whether individuals or parties act in a reasonably prudent manner under a given set of circumstances. **Liability** is based on cause. Causation in fact means that the injury can be traced back to the defendant's action (or lack of it). For proximate cause, the injury has to be reasonably foreseeable. Thus, a person cannot be held liable for unforeseen circumstances.

Acts of God

Individuals or parties generally are not held liable for injuries resulting from events determined to be "acts of God." There are many legal precedents for landowners using the act of God defense in cases where trees have fallen or broken and caused damage or personal injuries. Many times this defense has not been accepted. The reason lies in the definition of an "act of God." The courts have defined it a number of ways, but the principle remains the same. An act of God is an occurrence due to natural causes that could not have been prevented by ordinary skill and foresight. Thus, individuals will not escape liability for damages resulting from a fallen tree or branch if the defect that caused the failure was known of or should have been known of. Arborists have a duty of care to clients and employees and must exercise due diligence in inspecting and caring for the trees under their care. If a tree fails and causes injury, the act of God defense will not be applicable if it can be shown that the tree was structurally unsound or otherwise defective and should have been remedied or removed.

Arborists must understand the responsibility and liability that accompany tree risk assessment. Because arborists are considered experts in the care of trees, they can be held to a higher standard for inspecting and recognizing hazards, even if they were hired for other purposes than tree risk assessment. It is important to document any potential hazards that are discovered, to provide that information to the clients or property managers, and to maintain good records.

Chapter 12 Workbook

1. What are the three components of risk assessment?

    a.

    b.

    c.

2. By definition, if there is no ____________, a tree cannot pose a hazard.
3. When performing a tree assessment, it is important to develop and stick to a ___________ process.
4. Branches should be ____________ in diameter than their parent limbs or trunk.
5. ________________ stems have a high probability of failure.
6. Codominant stems often have _____________ ______ within the branch union, making them structurally unstable.
7. Trees that have grown in dense shade such as a forest tend to have long, straight stems with little ________.
8. Trees often produce ____________ ______, providing more stability to compensate for lean.
9. Cracks that start at a __________ ________ may be especially prone to failure.
10. ___________, even if it is compartmentalized, can be a significant risk factor for a tree.
11. Most experts agree that a threshold of __________ percent loss of support wood due to decay requires some action be taken.
12. A tree may appear to be solid, structurally sound, and may have a thick, green canopy yet can have significant _________ inside.
13. List five symptoms or signs of decay in a tree:

    a.

    b.

    c.

    d.

    e.

14. If decay is suspected at the root flare or in the major support roots, a root collar ____________ may be necessary to ascertain the extent.
15. Sometimes there are management options short of tree removal to abate the risk of failure. _____________ is the process of reducing the risk potential.
16. Current research indicates that filling cavities may do more _________ than good because decay can develop in the interface between filler and tree.
17. List four potential options for mitigating tree risks:

    a.

    b.

    c.

    d.

18. _______________ is the failure to exercise due care.

19. An _____ ___ _____ is an occurrence due to natural causes that could not have been prevented by ordinary skill and foresight.

20. Because arborists are considered to be experts in the care of trees, they can be held to a __________ _____________ for inspecting and recognizing defects.

CHALLENGE QUESTIONS

1. Develop sample evaluations for three trees that vary in condition, size, and location. Consider the target, the environment, and the potential for damage.

2. Why is cavity filling generally not recommended? Describe the possible consequences of cavity filling, taking into account the CODIT process.

3. What role do arborists play in recognition and assessment of risks when they are hired to prune the trees on a property?

SAMPLE TEST QUESTIONS

1. A necessary component in the existence of a tree hazard is
   a. a tree with a potential for failure
   b. an environment that may contribute to failure
   c. a person or object that may be injured or damaged
   d. all of the above

2. Codominant branches can pose a potential risk because they
   a. are excurrent by nature
   b. have a weak union
   c. develop reverse branch taper
   d. form natural lion tails

3. Following construction, forest trees on the edge of remaining stands are prone to failure due to
   a. losing the protection of the trees that used to surround them
   b. less trunk stability and poor taper
   c. increased exposure to the weather elements
   d. all of the above

4. Trees that lean because of ground failure or root injury
   a. have a high potential to fail and should be removed
   b. are less of a risk than those that lean due to phototropism
   c. are not a threat unless located at the edge of a wooded area
   d. are a risk only if they begin to grow in compensation for the lean

5. Arborists, as trained professionals in tree care,
   a. are required to have formal training in hazard assessment
   b. must be qualified to perform a root crown inspection
   c. will be held to a higher standard of duty than general citizens
   d. all of the above

Other Sources of Information

(See pages v–vi for complete bibliographic information.)

Matheny and Clark, 1994. *A Photographic Guide to the Evaluation of Hazard Trees in Urban Areas.*
Merullo and Valentine, 1992. *Arboriculture and the Law.*

CHAPTER 13
TREES AND CONSTRUCTION

CHAPTER 13 TREES AND CONSTRUCTION

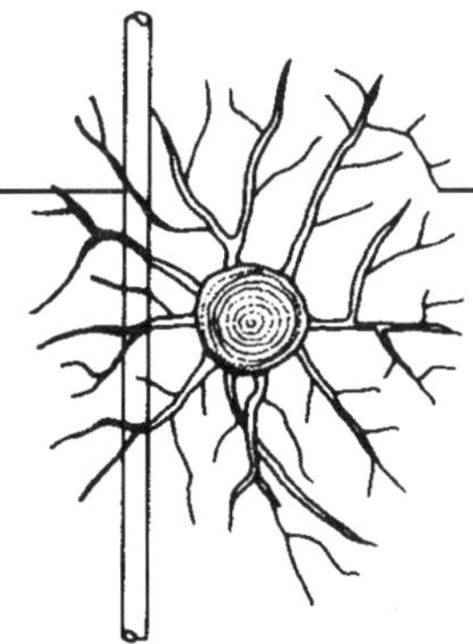

CHAPTER 13 TREES AND CONSTRUCTION

Objectives

1. Describe how trees can be injured or killed as the direct result of construction damage.
2. Discuss the importance of arborists' participation in the planning stages of development if trees are to be a part of the landscape.
3. Explain the steps that can be taken to preserve trees on a construction site.
4. Discuss some techniques that can be used to preserve trees when the soil grade must be changed.
5. Explain the limitations for treatment of trees that have been damaged by construction.

Key Terms

access route	bark tracing	soil compaction	tree island	tunneling
aeration system	barriers	specifications	tree well	vertical mulching
air excavator	radial trenching	terracing	trenching	

INTRODUCTION

Construction damage is one of the most common causes of tree death and decline in urban areas. As cities and suburbs expand, wooded lands are being developed into commercial and residential sites. Buildings are erected in the midst of trees to take advantage of the aesthetic value of the wooded lots. Unfortunately, if proper steps are not taken to ensure their survival, many of the trees will be lost in subsequent years.

It is possible to preserve trees on building sites if the right measures are taken. The most important step is to be sure that the professional arborist gets involved early—during the planning stage. An arborist can help decide which trees are suitable for saving and can work with the builder to protect trees throughout each phase of construction. Understanding how to minimize construction injury is based on knowledge of tree physiology and the components needed for tree health, as well as understanding construction practices.

HOW TREES ARE DAMAGED DURING CONSTRUCTION

The processes involved with construction can be deadly to nearby trees. Further, unless the damage is extreme, the trees may not die immediately, but could decline over several years. With this delay in symptom development, the loss of the tree may not be associated with the construction.

When construction equipment such as backhoes, bulldozers, and cranes are operated near trees, the trees are likely to be damaged. Branches can be broken or split. Often, tree trunks are seriously wounded. Pruning cuts made by untrained construction workers may form wounds that never close on a weakened tree (Figure 13.1).

The most serious damage to trees caused by construction is underground. The root system of a tree growing in a wooded area may spread a distance much greater than the height of the tree. Generally, the fine, absorbing roots are concentrated in the upper few inches of soil (Figure 13.2). These roots can easily

Figure 13.1 There are many ways trees can be damaged during construction. Successful tree preservation begins at the planning stage to avoid scenarios such as this one.

be damaged or killed by construction equipment. When tree roots are removed or damaged due to construction activity, the trees may show decline symptoms within a few months, or the symptoms may not appear for a few years. Symptoms may include small or yellow leaves, premature fall color, extensive watersprout development on the trunk and main limbs, dead twigs, and, eventually, major branches may die.

Physical Injury to Trunk and Crown

Construction equipment can injure the aboveground portion of a tree by breaking branches, tearing the bark, and wounding the trunk. These injuries are permanent, even if they are compartmentalized. If the injuries are extensive and the tree is stressed, the tree may never recover.

Cutting of Roots

The digging and **trenching** necessary to construct a building and install underground utilities will likely sever a portion of the roots of many trees in the area.

Figure 13.2 The roots of these trees have been exposed, showing just how shallow they are. Knowing this helps the arborist understand why compaction or grade increases can be so damaging to trees.

If the grade is lowered near a tree, a large percentage of the root system may be removed. With an understanding of where roots grow, it is easy to appreciate the potential for damage. The roots of a mature tree can extend far from the trunk of the tree. In fact, roots typically will be found growing a distance of one to three times the spread of the branches. The amount of damage a tree suffers from root loss depends, in part, on how close to the tree the cut is made. Severing one major root can cause the loss of 15 to 25 percent of the root system.

Another problem that may result from root loss due to digging and trenching is an increased potential for trees to fall over. Roots play a critical role in anchoring a tree. If the major support roots are cut on one or more sides of a tree, the tree may fall or blow over (Figures 13.3 and 13.4).

Soil Compaction

An ideal soil for root growth and development is about 50 percent pore space. These pores, the spaces

Figure 13.3 Even the construction of a path close to trees can sever a large percentage of the roots.

Figure 13.4 The installation of utilities in this trench followed the clearing of the site, grading, and sidewalk installation. The accumulated injuries will soon be apparent as the tree begins to decline.

Figure 13.5 This tree, located on a college campus, is in an advanced stage of decline from soil compaction, installation of the path, and construction injuries to the trunk.

between soil particles, are filled with water and air. **Soil compaction** can be devastating to trees. When soil is compacted, the pore space between soil particles is greatly reduced. This reduces oxygen availability to roots and causes accumulation of carbon dioxide and other gases. Root growth may be diminished, and the ability to absorb water and minerals decreased. Soil compaction reduces water infiltration and movement, and impairs drainage. The ability of roots to grow and expand into compacted soils is also reduced (Figure 13.5).

Smothering Roots by Adding Soil

Most people are surprised to learn that 90 percent of the fine roots that absorb water and minerals are in the upper few inches of soil. Roots require space, air, and water. Roots grow best where these requirements are met, which is usually very near the soil surface. Piling soil over the root system or increasing the grade smothers the roots. It takes only a few inches of added soil to kill a sensitive, mature tree. Even if the grade change is not in the immediate vicinity of the root zone, the water table or drainage pattern could be affected, which can adversely impact trees.

Exposure to the Elements

Trees in a forest grow as a community, protecting each other from the elements. The trees grow tall, with long, straight trunks and high canopies. Removal of neighboring trees, or opening the shared canopies of trees, exposes the remaining trees to sunlight and wind. The higher levels of sunlight may cause sunscald on the trunks and branches of trees with thin bark. Most forest trees have long, thin trunks, with little taper. If neighboring trees that helped provide protection for many years are removed, the remaining trees will be more prone to breaking from wind or ice loading (Figure 13.6).

To fully understand the effects of construction on trees requires comprehension of the biological needs and growth processes of trees, both as individual specimens and in groups or forests. Every aspect of tree health, growth, structure, and development can be affected. The impacts of construction are cumulative and can send a tree into a spiral of decline that might not even be recognized by the untrained eye until it is far too late. This fact underscores the need to involve a professional arborist early in the planning stages.

Figure 13.6 Homes are often constructed on wooded sites, and the lots can be valued at thousands of dollars more than sites without trees. However, unless measures are taken to conserve the trees, they are likely to die following construction.

PLANNING AND PRESERVATION

If trees are to be preserved on construction sites, their consideration cannot wait until the construction begins. In fact, the earlier in the process that an arborist becomes involved, the greater the chances of success.

Unfortunately, in the past, arborists were rarely involved in the development process. Arborists are now becoming a more integral part of the process, however. Arborists must learn to communicate and work with other professionals such as developers, planners, city engineers, and builders. Many levels of planning and development are involved with building a residential home; these levels are multiplied many times over for a complex city development. Standards and **specifications** exist for design and construction of buildings, installation of utilities, building roads and other paved surfaces, and every other aspect of development. Few professionals in these areas of specialization have any appreciation for the requirements of trees. If arborists can become a part of the development team, the needs of any trees existing on the site or to be planted later can be considered in the early stages of planning. If success is to be achieved, though, it will require the commitment of everyone involved, at each level of development.

One objective of development should be to satisfy all of the requirements of construction with minimal impact to trees that are to remain on site. The arborist must work with the builder and developers to determine which trees are to be preserved and which should be removed. It will probably not be practical, or even desirable, to save every tree. The decision must consider the species, size and maturity, location, and the condition of each tree. The arborist should evaluate each tree's condition and suitability for saving. Only healthy, structurally sound trees should be preserved. In addition, it is necessary to take into account the trees' tolerance for construction processes. The largest, most mature trees are not always the best choices to conserve. Younger, more vigorous trees can usually survive and better adapt to the stresses of construction. Try to maintain diversity of species and tree maturity.

An arborist should be involved in evaluating, selecting, and mapping trees worthy of consideration for preservation on a site. Trees should be considered along with all other landscape features (for example, wetlands, streams, and wildlife habitats) during the preliminary design phase. The arborist should prepare—often with the help of a landscape architect—a plan that details the location of trees and other features in relation to the buildings, roads, and infrastructure. This plan should show preservation zones and details of all mitigation and preservation measures for the project (Figure 13.7).

Sometimes design plans or construction procedures can be modified to better accommodate trees. Small changes in the placement or design of a building can make a great difference in whether a critical tree will survive. An alternative plan may be more friendly to the root system. For example, bridging over the roots may substitute for a conventional walkway. Or, instead of trenching beside a tree for utility installation, **tunneling** under the root system is much less damaging (Figure 13.8).

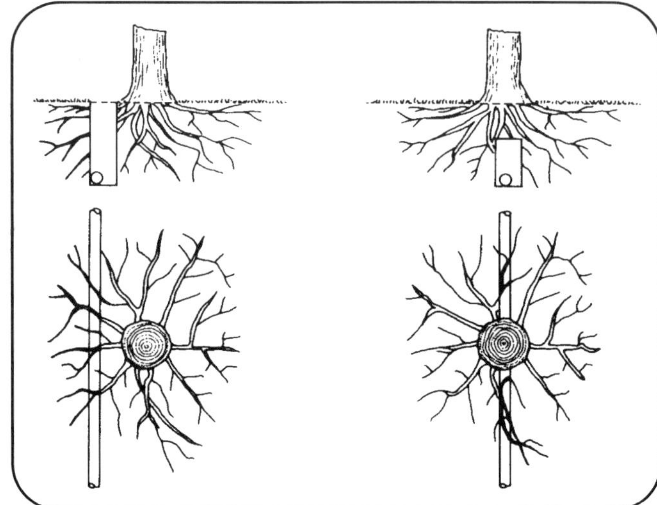

Figure 13.8 Excavation for utilities. Less damage is done to the root system if utilities are tunneled under a tree rather than across the roots.

Specifications

Get it in writing. All of the measures intended to protect trees must be written into construction specifications. The written specifications should detail exactly what can and cannot be done to and around the trees. Each subcontractor must understand the purpose of the **barriers**, limitations, and specified work zones. It is a good idea to post signs as a reminder.

Fines and penalties for violations should be built into the specifications. Not too surprisingly, subcontractors are much more likely to adhere to tree preservation clauses if their profit is at stake. The severity of the fines should be proportional to the potential damage to the trees and should increase for multiple infractions.

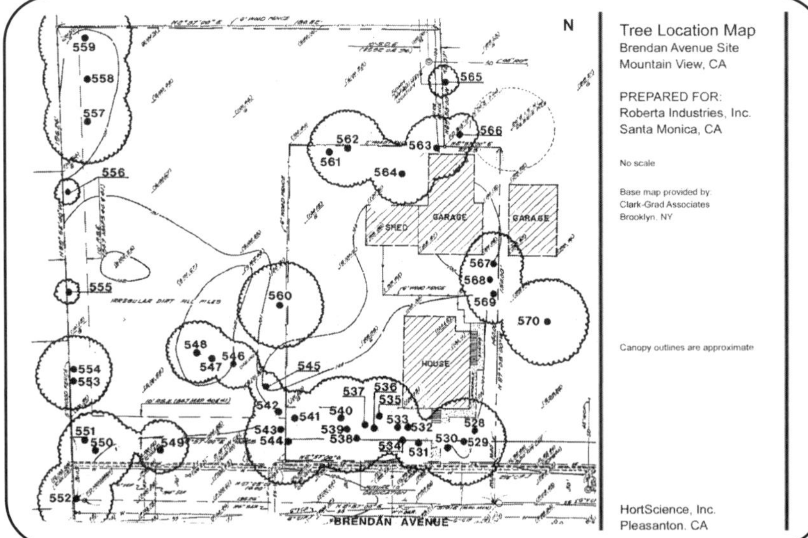

Figure 13.7 A tree survey map depicts the location of individual trees (often referencing tree tag numbers) and their canopy outlines. A topographic plan is generally used as a base map.

AVOIDING TREE DAMAGE DURING CONSTRUCTION

After all the plans have been made, the design plans finished, and the tree assessments completed, it will be time to implement measures for tree preservation on the construction site. Because arborists' ability to repair construction damage to trees is limited, it is vital that the trees be protected from injury.

Erecting Barriers

The single most important action that can be taken at the start of the construction process is to set up construction fences around all of the trees that are to remain. The fences should be placed as far out from the trunks of the trees as possible. As a general guideline, allow 1 foot of diameter from the trunk for each inch of trunk diameter. The intent is not merely to protect the aboveground portions of the trees but also the root systems. Remember that root systems can extend much farther out than the branches of trees (Figure 13.9).

Instruct construction personnel to keep the fenced area clear of building materials, waste, and excess soil. No digging, trenching, compaction, or other soil disturbance should be allowed in the fenced area.

Figure 13.9 These trees have been roped off to designate that they are to be preserved. Protective fences or barriers would be more effective—and should be placed much farther out from the trees to protect the root systems.

Limiting Access

If at all possible, it is best to allow only one **access route** on and off the property. All contractors must be instructed as to where they are permitted to drive and park their vehicles. Often this same access drive will later serve as the route for utility wires, waterlines, the driveway, or other paved surfaces.

Limit storage areas for equipment, soil, and construction materials. Specify areas for burning (if permitted), cement washout pits, and construction work zones. These areas should be away from protected trees.

Reducing Compaction

One technique to reduce compaction on construction sites is to spread a thick layer of mulch, such as wood chips readily available to arborists. The mulch should be 6 to12 inches deep. This has the effect of dispersing the weight of construction equipment. Additional weight dispersal can be obtained by placing large plywood sheets over the mulch. Construction mulching is a temporary measure, and the mulch cannot be left in place at that thickness for a prolonged period of time. The mulch must be removed carefully so as not to damage the trees in the process. An alternative is to spread the mulch, dispersing it to a layer of 2 to 4 inches in depth.

Changes in Grade

Grade changes can be devastating to trees, even if the change is not severe. If the grade must be lowered, the ability of a tree to survive depends on several factors. The most important is the amount of root system that will be cut away. Other important considerations are the tree species, the degree to which the grade must be lowered, the soil conditions, and whether the area is to be paved. If the grade must be lowered near a tree, **terracing** may help preserve some of the roots. The original grade should be maintained as far out from the trunk as possible. Terraces can be formed to lower the grade in steps. At each level, the roots should be severed cleanly and kept moist. If walls are installed at the level changes, they must allow for drainage. Sand is sometimes used as a backfill behind the walls to encourage new root growth.

If the grade must be lowered completely around a tree, a **tree island** can be constructed. Obviously, the greater the percentage of root system that remains at the original grade, the greater the chance of tree survival. The techniques used for creating a tree island are similar to those used in terracing (Figure 13.10).

If the grade is raised, the roots may be suffocated. As little as 4 inches of soil placed over the root system can kill some species. Increases in grade may require major steps to protect the tree. Sometimes **aeration systems** are installed to preserve trees, and designs of complex aeration systems and **tree wells** are described in various texts. There is little research, however, to confirm the value of these systems. Very large tree wells may keep soil fill far from tree trunks and minimize the percentage of the root system covered. Small-diameter wells built around the trunks of trees are rarely adequate to provide enough oxygen to keep trees alive.

Some specifications call for use of gravel or stone below the fill dirt to increase water and oxygen penetration. However, if soil is placed over gravel, water will not drain out of the soil layer until the saturation point is reached. If fill soil is placed in direct contact with the original grade soil, water penetration will be improved, and conditions will be more favorable for root penetration and development in the added fill soil.

Measures taken to preserve trees on a construction site, such as tree well construction, can be very expensive and involved, and there is no guarantee of success. Consideration must be given to the tree size and species, drainage pattern, soil conditions, depth of fill, and future irrigation and maintenance plans before a well is constructed. Recommendations should be based on realistic expectations for tree survival.

Maintaining Good Communications

It is important to work together as a team. The best-laid plans between arborist and builder can be destroyed by one uninformed subcontractor. The consulting arborist should visit the site at least once a day if possible. Vigilance will pay off as workers learn to take your wishes seriously. It is a good idea to take photos at every stage of construction. If any infraction of the specifications does occur, it may be important to link cause and effect.

Figure 13.10 Tree islands can be constructed to preserve trees on a construction site.

Final Stages

It is not unusual to go to great lengths to preserve trees during construction, only to have them injured during landscaping. Installing irrigation systems and rototilling planting beds are two ways the root systems of trees can be damaged. Careful planning and communicating with landscape designers and contractors is just as important as avoiding tree damage during construction.

Post-Construction Tree Maintenance

Trees on a construction site require several years to adjust to the injury and environmental changes that occur during construction. Stressed trees are more prone to health problems such as disease and insect infestations. Plan for continued maintenance, and monitor the trees' health, structure, and overall vitality.

Despite the best intentions and most stringent tree protection measures, some trees may still be injured by the construction process. There are remedial treatments that an arborist can use to help reduce stress and improve the growing conditions around the trees, but these treatments are few. Unfortunately, the ability to restore trees damaged by construction is very limited, and this is why the emphasis should be placed on protection.

TREATMENT OF TREES DAMAGED BY CONSTRUCTION

Inspection and Assessment

Because construction damage can affect the structure and stability of a tree, it is important to check for potential hazards. This may involve a visual inspection, or instruments may be used to check for the presence of decay. Sometimes a hazard can be reduced or eliminated by removing an unsafe limb, pruning to reduce weight, or installing cables or braces to provide structural support. An often-overlooked method of reducing hazards is to move objects that could be hit, or to limit access to the hazardous area. If there is doubt about the structural integrity of a tree or the risk cannot be adequately reduced, the tree should be removed. Although the goal is to preserve trees whenever possible, that goal must not supersede any question of safety.

Treating Trunk and Crown Injuries

Pruning

Branches that are split, torn, or broken will probably need to be removed. In addition, any dead, diseased, or rubbing limbs should be removed. Sometimes it is necessary to remove some lower limbs to raise the canopy of a tree and provide clearance below. It is best to postpone other maintenance pruning for a few years.

Old recommendations suggest that tree canopies should be thinned or topped to compensate for root loss. There is no conclusive research to support this practice. Thinning the crown can reduce the tree's food-making capability and may stress the tree further. It is better to limit pruning in the first few years to hazard reduction and the removal of deadwood, especially in large, mature trees.

Cabling and Bracing

Trees growing in wooded areas are usually not a threat to people or structures. Trees close to houses or other buildings must be maintained to keep them from injuring people or damaging property. If branches or tree trunks need additional support, risk of failure may be reduced with the installation of cables or bracing rods. If cables or braces are installed, however, they must be inspected regularly. The amount of added security offered by the installation of support hardware is limited. Not all weak limbs are candidates for these measures.

Repairing Damaged Bark and Trunk Wounds

Often the bark may be damaged along the trunk or major limbs. If this happens, the loose bark should be carefully removed. Jagged edges can be cut away with a sharp knife, taking care not to cut into living tissues. This procedure is called **bark tracing.**

Wound Dressings

Wound dressings were once thought to accelerate wound closure and reduce decay. Research has not substantiated this, however. Some studies have shown beneficial effects in specific cases in reducing borer attack, oak wilt infection, or control of sprout production or mistletoe. However, wound dressings are primarily used for cosmetic purposes, and are neither required nor recommended in most cases. If a dressing must be applied, only a light coating of a nonphytotoxic material should be used.

Irrigation and Drainage

One of the most important tree maintenance procedures following construction damage is to maintain an adequate, but not excessive, supply of water to the root zone. If there is a drainage problem, the trees will decline rapidly. This problem must be corrected if the trees are to be saved. If soil drainage is good, be sure to keep the trees well watered, especially during the dry summer months. A long, slow soak over the entire root zone is the preferred method of watering. Frequent, shallow waterings should be avoided, and water should not be directed at or near the trunks of trees.

Mulching

One of the simplest and least expensive things that can be done for trees may also be one of the most effective. Applying a 2- to 4-inch layer of an organic mulch such as wood chips, shredded bark, or pine needles over the root system of a tree can enhance root growth. The mulch helps condition the soil, moderates soil temperatures, maintains moisture, and reduces competition from weeds and grass. The mulch should extend as far out from the tree as is practical for the landscape site. When it comes to mulch, deeper is not better, and piling it up against the trunk can lead to disease problems.

Aeration of the Root Zone

Compaction of the soil and increases in grade both deplete the oxygen supply to tree roots. If soil aeration can be improved, root growth and water uptake can be enhanced. Vertical mulching and radial trenching are techniques that may improve conditions for root growth. If construction-damaged trees are to survive the injuries and stresses they have suffered, they must replace the roots that have been lost. Research to date on improving aeration has been limited, and there is still some question how successful attempts at improving aeration have been (Figure 13.11).

Figure 13.11 This hackberry (*Celtis occidentalis*) suffered significant root loss from construction. Efforts were made to preserve the tree, and arborists have removed some of the deadwood that resulted from construction. The root area has been aerated, and the health of the tree is being monitored.

Drilling Holes/Vertical Mulching

The most common method of aeration of the root zone involves drilling holes in the ground. Holes are usually 2 to 4 inches in diameter and are made about 1 to 3 feet on center, throughout the root zone of the tree. The depth should be at least 12 inches but may need to be deeper if the soil grade has been raised. Sometimes the holes are filled with organic material such as compost, peat moss, or other materials that maintain aeration and support root growth. This is called **vertical mulching**.

Radial Trenching

Another method of aeration is **radial trenching**. Trenches are dug in a radial pattern throughout the root zone. The trenches appear similar to the spokes of a wagon wheel. If a mechanical trencher is used, it is important to begin the trenches 4 to 8 feet from the trunk of the tree to avoid cutting any major support roots. The trenches should extend at least as far as the drip line of the tree. Use of an **air excavator** has proved very effective for this purpose. While mechanical trench digging sacrifices some roots, air excavation devices do little damage to roots, even if trenching is not radial. If the primary goal is to reduce compaction, the trenches should be about 1 foot in depth. They may need to be deeper if the soil grade has been raised.

The narrow trenches are backfilled with native soil and compost. Root growth will be greater in the trenched area than in the surrounding soil. This can give the tree the added boost it needs to adapt to the compacted soil or new grade.

What About Fertilization?

Most experts recommend that trees not be fertilized the first year after construction damage because water and mineral uptake may be reduced due to root damage. If fertilizer is applied, however, slowly soluble mineral sources should be used to minimize the potential for root injury. All fertilizers release mineral salts in soil water, which are absorbed by roots. Soluble sources of fertilizer can lead to excessive soil salts, which can draw water out of the roots and into the soil. In addition, soluble nitrogen fertilization may stimulate top growth at the expense of root growth, particularly if root growth is suppressed. It is a common misconception that applying fertilizer gives a stressed tree a much-needed shot in the arm. Fertilization should be based on the nutritional needs of trees. If essential elements are deficient, supplemental fertilization may be indicated. It is advisable to keep application rates low, however, until the root system has had time to adjust.

Monitoring for Decline and Hazards

Despite everybody's best efforts, some trees may be lost as a result of land development and construction activities. If a tree dies as a result of root damage, it may be an immediate hazard and may require immediate removal. Trees may appear to recover from construction damage but may have extensive root rot. Arborists must inspect and monitor trees that have suffered construction stresses for signs of possible hazards. Look for cracks in the trunk, split or broken branches, and dead limbs. Watch for indications of internal decay such as cavities, carpenter ants, soft wood, and mushroomlike structures growing on the trunk, root collar, or along the major roots. Trees should also be inspected for signs of insects or diseases. Stressed trees are more prone to attack by certain pests.

Chapter 13 Workbook

1. Name five ways that trees can be adversely affected by construction.
   a.
   b.
   c.
   d.
   e.
2. When soil is compacted, the ________ ___________ between soil particles is reduced.
3. Two detrimental effects of soil compaction on tree roots are
   a.
   b.
4. If the soil grade on a construction site is ____________, a large percentage of a tree's root system might be removed. If the grade is ______________, the tree's roots might be suffocated.
5. A technique used to reduce soil compaction around trees on a construction site is to spread a temporary, thick layer of ___________.
6. True/False—The overriding objective of an arborist involved in a development project is to save every tree on the site.
7. True/False—It is better to tunnel directly under a tree than to cut directly across the root system of a tree when excavating for utility lines.
8. An important action that should be taken at the start of a construction project is to erect _____________ around all of the trees that are to remain.
9. Carefully cutting away loose, damaged bark is called _______ _____________.
10. Soils that have been compacted or raised in grade are good candidates for soil _______________.
11. True/False—There is far more that an arborist can do to treat trees that have been damaged by construction than to prevent the damage.
12. _________________ is a technique that may be employed to lower the soil grade in steps.

CHALLENGE QUESTIONS

1. Write a sample set of specifications to preserve trees on a construction site. Include specifications for protecting trees before and during construction, as well as maintenance instructions to help maintain the health and vigor of trees.
2. What actions can be taken if a tree is damaged by construction in violation of the written specifications?

3. Why may tree death and decline due to construction occur several years after construction is complete? What are some of the signs and symptoms of construction damage that an arborist can look for following construction?

SAMPLE TEST QUESTIONS

1. When soils are compacted by construction equipment, trees usually decline because
   a. oxygen availability is reduced
   b. the ability of the roots to absorb water and minerals decreases
   c. root growth and expansion may be diminished
   d. all of the above
2. Arborists should be involved early in the construction planning process because
   a. tree preservation measures should be in the specifications
   b. once construction has begun, it may be too late to save the trees
   c. there is often little arborists can do to treat construction damage
   d. all of the above
3. A measure that can be taken to minimize compaction on a construction site is
   a. water the site thoroughly before equipment is brought in
   b. permanently raise the soil grade to protect tree roots
   c. spread a temporary, thick layer of mulch over the site
   d. root prune the trees in advance
4. If a significant portion of a tree's root system has been removed during building construction, a step that will help preserve the tree is
   a. pruning one-third of the crown to compensate for root loss
   b. a surface application of 10 pounds of soluble nitrogen per 1,000 square feet
   c. construction of a tree well
   d. none of the above
5. Digging trenches in a wheel-spoke pattern and backfilling with organic matter or a more porous soil is called
   a. radial aeration
   b. tunnel aeration
   c. soil fracturing
   d. vertical radiation

Other Sources of Information

(See pages v–vi for complete bibliographic information.)

Matheny and Clark, 1998. *Trees and Development: A Technical Guide to Preservation of Trees During Land Development.*
Watson. *Root Injury and Tree Health.*

CHAPTER 14
SAFETY

CHAPTER 14 SAFETY

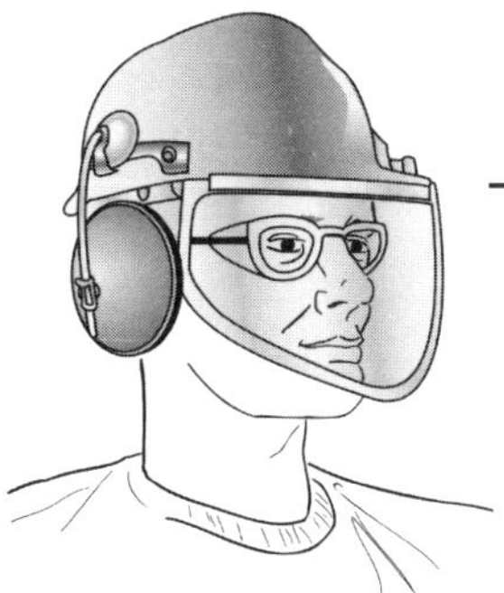

CHAPTER 14
SAFETY

Objectives

1. Become familiar with safety standards for tree care operations.
2. Describe the potential hazards associated with working in proximity to electrical conductors and the fundamental measures necessary to avoid direct or indirect contact.
3. Explain the proper procedures for operating chain saws and chippers and the personal protective equipment required for their operation.
4. Read and understand the first-aid procedures and understand the steps that must be followed in the event of an emergency.

Key Terms

ANSI Z133.1
aerial rescue
approved
back cut
barber chair
Canadian Standards Association (CSA)
cardiopulmonary resuscitation (CPR)
chaps
command and response system
conventional notch
direct contact
drop zone
electrical conductor
emergency response
first aid
hinge
Humboldt notch
indirect contact
job briefing
kickback
kickback quadrant
landing zone
leg protection
open-face notch
Occupational Safety and Health Act (OSHA) (OHSA in Canada)
reactive forces
shall
should
tagline
work plan

INTRODUCTION

Working in and around trees can be a very hazardous profession if proper care and safety measures are not followed. Safety must always be the first concern. Safety is more than using special equipment, wearing appropriate gear, or attending occasional meetings. Safety is an attitude. It is an ongoing commitment at every level. Safety requires a conscious recognition of potential hazards and the development of a program designed to prevent accidents. Safety precautions must be built into every task performed by tree workers. A small investment of time in safety training can save a great deal in down time, insurance, injuries, and damages.

LAWS AND REGULATIONS

In the United States, private employers are subject to the laws of the **Occupational Safety and Health Act**. The act is administered by the Occupational Safety and Health Administration **(OSHA)**. In Canada, it is the Occupational Health and Safety Act (**OHSA,** administered by CanOSH). The purpose is to reduce occupational injury, illness, and death through the establishment and enforcement of safety standards and regulations, and the provision for mandatory training. Many states also have occupational safety and health laws.

OSHA has established many regulations that affect arboricultural work. For example, OSHA regulates tree work in the vicinity of electrical conductors (1910.269). Most OSHA regulations are general in nature and govern multiple professions.

ANSI Z133.1 is a set of standards for arboricultural operations approved by the American National Standards Institute (ANSI). It is intended to provide safety standards for workers engaged in pruning, repairing, maintaining, or removing trees or cutting brush. Canadians must comply with the standards established by the **Canadian Standards Association (CSA)**.

ANSI standards are the recognized safety standards for tree care in the United States. They are developed and updated by a committee of tree care professionals, including representatives of the International Society of Arboriculture and the National Arborist Association. It is important for all tree care workers to be familiar with and comply with the ANSI Z133.1 standards and all applicable OSHA regulations. In Canada, regulations governing tree care operations are very similar and sometimes are more stringent.

Certain terminology is consistent through most safety regulations in the United States and Canada. **Approved** means acceptable to the federal, state, provincial, or local enforcing authority having jurisdiction. In many cases, the word "approved" appears pertaining to certain equipment. The ANSI Z133.1 standards often reference other pertinent standards that regulate certain types of equipment. Two additional terms are important to know. **Shall** denotes a mandatory requirement; **should** denotes an advisory recommendation.

This text cannot reference all of the pertinent standards and regulations for tree work performed in the United States and Canada. Tree workers are reminded and encouraged to become familiar with these standards. Also, the content of these standards takes precedence over this and other educational texts. ***Because of inconsistencies between standards, and for the sake of readability, the use of the terms "shall," "should," and "must" in this text does not necessarily parallel the requirements of ANSI, OSHA, or CSA standards.***

PERSONAL PROTECTIVE EQUIPMENT (PPE)

Tree workers should wear clothing and footwear appropriate for the work conditions and weather. Fabrics should be durable yet allow for free movement. Loose-fitting clothing may catch in machinery and become a hazard. Jewelry should not be worn because it may catch in equipment.

All tree workers must wear head protection (hard hats or approved helmets) (Figure 14.1). Hard hats must comply with federal impact and penetration requirements. Protective headgear shall conform to the applicable provisions of ANSI Z89.1. Class-E hard hats shall be worn when working in proximity to an electrical conductor.

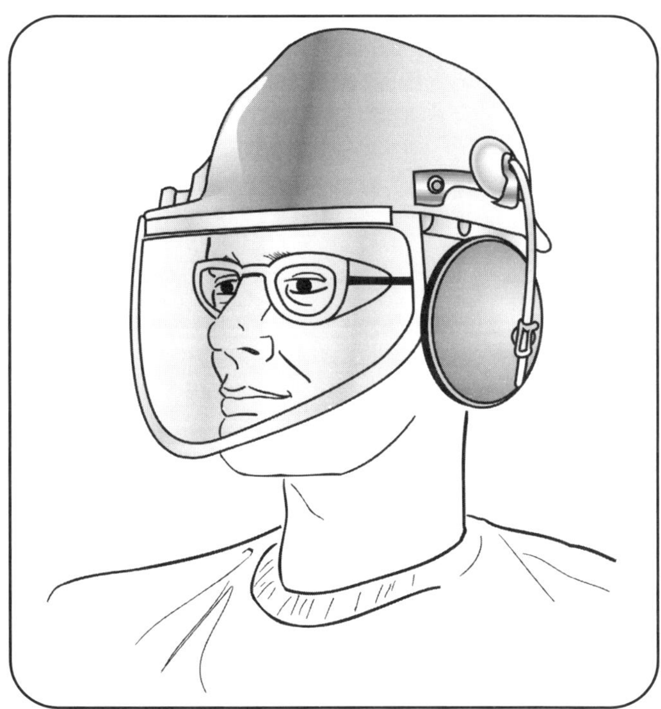

Figure 14.1 Head protection is required for tree work. This hard hat has an incorporated face shield, which provides added protection but cannot be considered a substitute for eye protection. Safety glasses are required.

Eye protection must also be worn when performing tree work. If a worker is poked in the eye by even a small twig, irreparable damage may be done to the eye. Sawdust and wood chips from chain saws and chippers pose a significant threat to a worker's eyes. Tree climbers should wear protective safety glasses or goggles. Some are available with UV protection as well. Some workers also prefer to wear face shields, such as those attached to some hard hats. Face shields offer additional protection to the face but must not be considered a substitute for safety glasses.

It has been demonstrated that prolonged exposure to the noise of chain saws and brush chippers can cause permanent hearing loss. Hearing protection is required for workers who are exposed to loud equipment for prolonged periods of time (90 dB on a time-weighted average for an eight-hour day). Because tree workers use chain saws and chippers frequently, it is recommended that they wear hearing protection whenever operating them. Workers may choose to use either ear-muff– or ear-plug–type hearing protection equipment, provided those devices chosen meet appropriate standards (Figure 14.2).

Regulations requiring the use of gloves in tree care operations vary. Some regulations require gloves for tree workers. However, ANSI Z133.1 does not make that requirement. Gloves are strongly

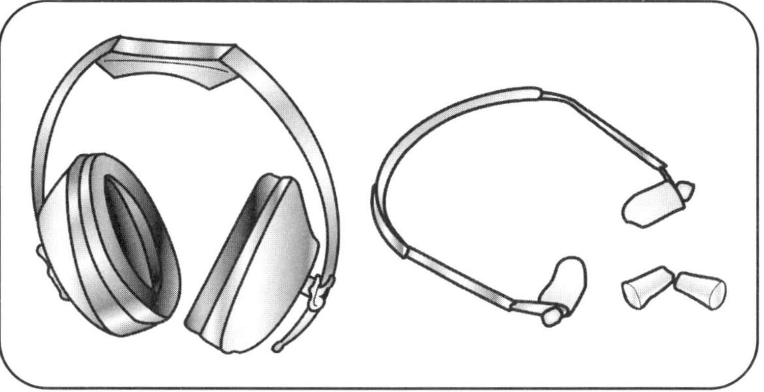

Figure 14.2 Hearing protection is required for workers who are exposed to loud equipment, such as chain saws and chippers, for prolonged periods of time.

recommended for certain operations such as sharpening chain saws and chipping brush. Workers must not wear gauntlet-type gloves while chipping brush because these gloves may get caught up in the brush. If gloves are to be worn when climbing trees, they should be a type that provides a good grip to prevent the climber's hands from slipping.

All tree workers should wear sturdy work boots to provide good support and protection for the feet. Many styles and types of approved work boots are available. Some newer designs have chain saw protection. If climbing spurs are used frequently, climbers might choose boots with a deep, square heel to brace the stirrup of the climbing spur. Other boots are designed with flat soles to facilitate footlocking.

When chain saws are used on the ground, **leg protection** must be worn (Figure 14.3). Leg protection may be in the form of **chaps** or chain saw pants. The fabric of these leg protection devices is designed to jam and slow the cutters of the chain saw if contact is made. Leg protection has been shown to reduce the frequency and severity of chain saw injuries. Newer designs of chain saw pants are lightweight and less bulky than early models. Some companies now require climbers to wear chain saw pants when using chain saws in trees. Other work clothing has been developed that has chain saw protection incorporated into the fabric. Examples include shirts, jackets, and bib-style pants.

A relatively new article of personal protective equipment available to tree workers is the back support. Some saddles are now available with wider back pads for added support and comfort. Other back-support devices can be worn while working on the ground. The idea is to provide support to the lower back to help reduce the probability of back injuries. Keep in mind that back-support belts do not substitute for proper lifting techniques. Prolonged use of back-support devices may actually allow the back muscles to weaken, making the user more susceptible to injury when the belt is not worn.

GOOD COMMUNICATION

Good communication among workers is an integral part of working safely. Workers must coordinate operations on the ground and aloft in the trees, and there is little margin for error. Each worker on the crew must always be aware of what the others are doing, and each must take measures to prevent accidents. There must be a clear and efficient means of communication between climbers and ground workers so that each knows when it is safe for a ground worker to enter the **landing zone** or **drop zone**, where branches are being dropped or lowered to the ground. The voice **command and response system** ensures that warning signals are heard, acknowledged, and acted upon. The climber warns, "Stand clear," but does not proceed until hearing the acknowledgment, "All

Figure 14.3 Leg protection must be worn by workers operating chain saws on the ground.

clear." There are times when it may be difficult for workers to hear each other. In these cases, hand signals can be used as well.

Each job should begin with a **job briefing,** which coordinates the activities of every worker. The job briefing summarizes what has to be done and who will be doing each task, the potential hazards and how to prevent or minimize them, and what special PPE may be required. The on-site supervisor should formulate and communicate the **work plan**. There should be no question about assignments, so that the work is well coordinated. Teamwork is essential on a tree crew.

GENERAL SAFETY

General safety at the work site begins with proper training (Figure 14.4). All workers must be adequately trained for their work requirements. Workers should be aware of all applicable safety regulations. Safe work procedures must be established. Employers shall instruct their employees in the proper use of all equipment and require that all safe working practices be followed.

It is recommended that all tree workers receive training in **first aid** and **cardiopulmonary resuscitation (CPR)**. In addition, an approved and adequately stocked first-aid kit must be provided on each truck (Figure 14.5). All employees should be instructed in the use of first-aid kits and in emergency procedures. Emergency phone numbers should be posted in the truck.

All crew members should be trained in **emergency response** and **aerial rescue** procedures. Each individual must know what to do in an emergency situation. All climbers should be trained and capable of carrying out an aerial rescue. Whenever there is a climber in a tree, there should be a second worker on site who is capable of performing an aerial rescue if necessary. Many companies practice aerial rescue procedures on a regular basis. Practice will improve efficiency and skills and will reduce the chance of panic and a second accident in the event of an actual emergency.

Figure 14.4 Proper training is an integral part of safe work operations.

Figure 14.5 An approved and adequately stocked first-aid kit must be provided on each truck. Workers should receive training in first-aid and emergency response procedures.

All employees should be instructed in the identification of common poisonous plants such as poison ivy, poison oak, and poison sumac. Training should include preventive measures as well as treatment following exposure. Workers should also be trained in techniques for dealing with stinging or biting insects or with vertebrates that could be encountered in trees (Figure 14.6).

Trucks should be equipped with a fire extinguisher, and all employees should be trained in its use. Gasoline-powered equipment shall be refueled only after the engine has been stopped. Any spilled fuel should be removed before starting. Do not start or operate the equipment within 10 feet (3 meters) of the refueling site. Smoking shall be prohibited when handling or working around any flammable liquid. Flammable liquids must be stored, handled, and dispensed from approved safety containers.

A very important safety consideration is traffic control. Effective means for control of pedestrian and vehicular traffic shall be instituted on every job site. This may include safety cones, warning signs, barriers, and flags. It is the legal responsibility of work

Figure 14.6 All workers should be instructed in the identification of poisonous plants such as poison ivy.

crews to "secure the work zone" to make sure that no individuals or vehicles pass under trees where tree work is in progress and to ensure the safety of the workers. Traffic control devices used in tree operations shall conform to applicable federal, state, or provincial regulations. Traffic control procedures should follow Department of Transportation standards and guidelines.

ELECTRICAL HAZARDS

Before performing tree work on any site, an inspection must be made to determine whether any electrical hazards exist. An electrical hazard exists when, in the course of climbing or working, the tree worker or any conductive object he or she contacts, comes within 10 feet of an energized conductor. All tree workers should receive adequate and documented training in electrical hazard safety awareness. Workers must receive proper training in electrical hazard tree work procedures to perform tree work in proximity to electrical conductors. Employers are required to certify this training. **Electrical conductor** is defined as any overhead or underground electrical device, including communication wires and cables, power lines, and related components and facilities. All such lines and cables shall be considered energized with potentially fatal voltages.

Every tree worker shall be instructed that a **direct contact** is made when any part of the body contacts an energized conductor or other energized electrical fixture or apparatus (Figure 14.7). An **indirect contact** is made when any part of the body touches any conductive object in contact with an energized conductor (Figure 14.8). An indirect contact can be made through conductive tools, tree branches, trucks, equipment, or other conductive objects, or as a result of communication wires or cables, fences, or guy wires becoming energized.

Electric shock occurs when a tree worker, by either direct or indirect contact with an energized conductor,

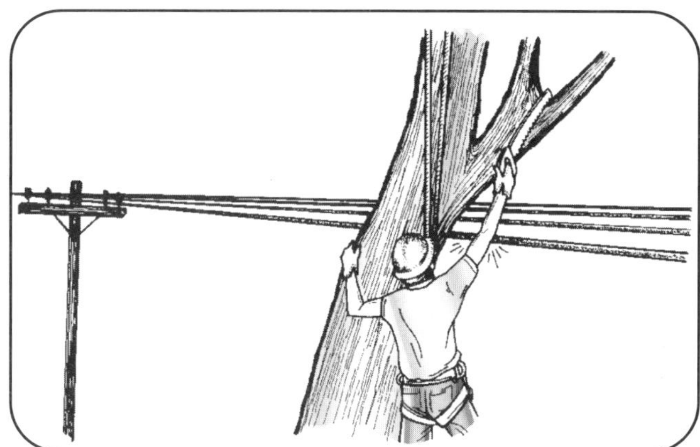

Figure 14.7 Direct contact is made when any part of the body contacts an energized conductor.

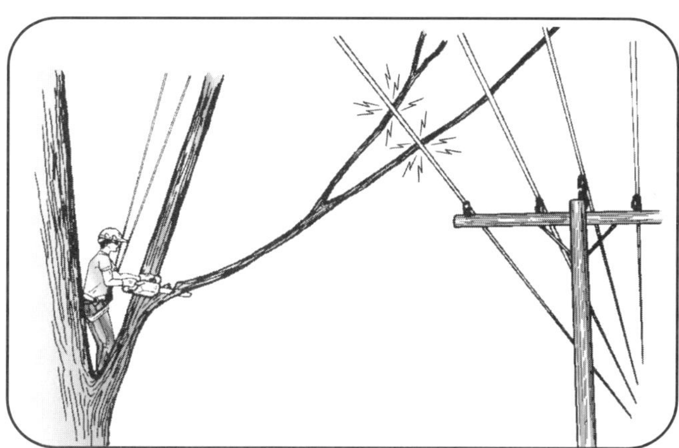

Figure 14.8 Indirect contact is made when any part of the body touches a conductive object in contact with an energized conductor.

energized tree limb, tool, equipment, or other object, provides a path for the flow of electricity from a conductor to a grounded object, or to the ground itself. Simultaneous contact with two energized conductors also causes electric shock that may result in serious or fatal injury.

Footwear, including those with electrical resistant soles and lineman's overshoes, shall not be considered as providing any measure of protection from electrical hazards. Rubber gloves, with or without leather or other protective covering, shall not be considered as providing any measure of protection from electrical hazards.

Electrical tools (except those with a self-contained power source) shall never be used in trees near an energized electrical conductor when there is a possibility of the power cord contacting the conductor. Tool operators should use tools in accordance with the manufacturers' instructions. When tools are used aloft, an independent line should support the electrical tool. Operators should prevent cords from becoming entangled or coming in contact with water.

CHAIN SAW SAFETY

The chain saw has been ranked as one of the most dangerous pieces of equipment in the industry today. It is also one of the most commonly used. Using a chain saw to prune or remove a tree can greatly reduce the time and effort involved, but care must be taken to ensure the safety of the climber and the ground workers. Careless use of a chain saw in a tree can cause considerable damage to the tree and serious danger to the workers. Safe operation on the ground and in the tree requires proper training and adherence to safety procedures.

Chain saw operators should follow the manufacturer's operating and maintenance instructions. Personal protective equipment that must be worn by chain saw operators includes a hard hat, work boots, eye protection, hearing protection, and, when on the ground, leg protection.

The operator should have secure footing when starting the saw and should ensure that the immediate area is clear of debris. The chain brake should always be engaged before the chain saw is started. Larger saws should be started on the ground or otherwise firmly supported. The saw should be gripped with both hands. A chain saw should never be operated with one hand while on the ground or aloft in a tree—both handles should be gripped firmly with the thumb wrapped around the handle. Chain saws are designed to be operated on the right side of the body. The back of the saw may be braced against the operator's right leg. Keeping the chain saw engine close to the body increases control and reduces operator fatigue.

The chain saw operator should be aware of the presence and activity of other workers in the vicinity. The chain saw operator should never be approached from the rear. If two workers are operating saws simultaneously, they should be at least 10 feet apart. The chain brake must be engaged if the operator takes one hand off the running saw or takes more than two steps. If the operator moves between cuts, the chain brake should be engaged and the operator's hand should be off the throttle lever. The engine must be stopped for all cleaning, refueling, and adjustments except where the manufacturer's procedures require otherwise.

Chain saw operators should understand the **reactive forces** of the saw. When cutting with the bottom of the bar, the saw has a tendency to pull into the cut. When cutting with the top of the bar, the saw pushes back toward the operator (Figure 14.9). A common cause of chain saw injury is **kickback**. Kickback occurs when the upper portion of the tip of the guide bar (**kickback quadrant**) contacts a log or other object (Figure 14.10). Operators should always be aware of where the tip of the guide bar is and should prevent the upper quadrant of the guide bar tip from coming in contact with objects (Figure 14.11). A firm grip should be maintained with the thumb wrapped around the forward handle. Kickback occurs at a speed many times faster than a human can react. The operator must be positioned such that the saw could not hit him or her if it were to kick back. Operators should avoid operating a chain saw above shoulder level.

Climbers must take extra precautions when using a chain saw aloft in a tree. Only experienced climbers with proper chain saw training should use a chain saw in a tree. The climber should use a second means

Figure 14.9 Workers must be aware of the forces acting on a tree or log under tension, as well as the reactive forces of the saw.

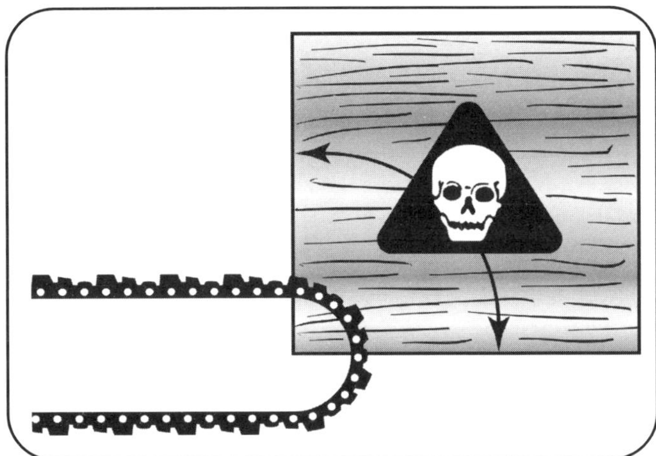

Figure 14.10 The upper tip of the guide bar is sometimes called the kickback quadrant.

Figure 14.11 Saw operators should be aware of where the tip of the guide bar is at all times and should prevent the upper tip from coming in contact with objects.

of securing, such as a safety lanyard, before making a chain saw cut. Stable footing and work positioning is important to maintain control. All chain saw safety procedures apply, including the two-hand rule and avoiding making cuts above shoulder level, when practical. Chain saws weighing more than 15 pounds should be supported by a separate line.

Work positioning is an important safety consideration when using a chain saw in a tree. Climbers must be sure they are in a safe, stable position before cutting. This means avoiding being in a position that could cause the climber to be injured in the follow-through from the cut, or if the saw kicks back. Climbers must also be aware of the position of the climbing line, safety lanyard, and all other ropes to avoid accidental cutting. Finally, it is important to avoid being in a position of being struck by the limb that is cut.

TREE FELLING AND REMOVAL

Before beginning any removal operation, the crew must consider all of the site and tree conditions that may affect the procedures used. This includes the surrounding obstacles and terrain; wind direction and magnitude; and the shape, flaws, and lean of the tree. If the structure of the tree appears to be compromised in any way, further investigation may be necessary before a plan for removal can be formulated. Once a work plan is formulated, workers should be positioned so as to avoid accidents. Many times, in the situations under which arborists work, trees must be rigged for removal. Other times, they can be felled using a **tagline** (pull line) for control. Workers who are not directly involved in the removal operation should be two tree lengths away from the tree being felled. Those involved must have a planned escape route. The preferable escape route for the chain saw operator in a felling operation is 45 degrees on either side of a line drawn opposite the intended direction of fall. As soon as the tree starts to fall, the feller should move away in this direction.

Felling notches and back cuts must be made high enough above ground level to enable the chain saw operator to begin the cut safely, control the saw, and have freedom of movement for escape. A number of notch configurations can be used for felling trees, including an **open-face notch**, a **conventional notch**, and a **Humboldt notch** (Figure 14.12). Commonly, a 45-degree notch is used. However, an open-face notch of 70 degrees or more allows the **hinge** to control the tree longer. The depth of the notch should not exceed one-third of the diameter of the tree. If possible, the notch should not be placed in the vicinity of cracks or decay because it is important to have solid fiber to form the hinge. It is very easy to bypass the apex of the notch when making the cuts. Doing so cuts into the crucial fibers of the hinge and must therefore be avoided.

The hinge is critical in controlling the direction of fall (Figure 14.13). If the hinge is the proper thickness, then the wood fibers should break when the face notch closes. The rule of thumb for felling trees is to allow a hinge with a thickness of 10 percent of the tree's diameter (Figure 14.14). Flexibility in this guideline is in order. For example, when cutting

conventional notch

Humboldt notch

open-face notch

Figure 14.12 Comparison of three notch configurations.

short sections, a 10 percent hinge may be too much for the climber to break off with limited leverage. For larger trees, a hinge of less than 10 percent may be better. Avoid cutting into the hinge when making the back cut.

The traditional, straight **back cut** is made from the back of the tree toward the notch. The hinge is formed as the back cut approaches the notch. It is easy to cut through the hinge while making the back cut, especially if the saw operator is looking toward the top of the tree. Most texts recommend making the back cut slightly higher than the apex of the notch to reduce the tendency of the tree to kick back toward the operator when the hinge breaks. This is important when using a conventional (45-degree) or Humboldt notch. When an open-face notch is used, the back cut can be made level with the apex of the notch.

Sometimes, if the tree is leaning in the direction of fall or has internal faults, the tree can split upward from the back cut. This is called a **barber chair**, and it can be very dangerous. The split trunk can hit the tree worker if the worker is directly behind the tree. The worker should be conscious of this possibility and try to remain to one side of the tree while making the back cut. No other workers should be in the area directly behind the tree. The worker making the back cut should plan an escape route at a 45-degree angle to either side of a line drawn opposite the intended direction of fall. Training in techniques for felling trees with lean or with flaws can reduce the possibility of a barber chair occurring.

It is a good idea to have felling wedges ready when making the back cut. Wedges can be driven into the back cut to prevent the tree from closing on the back cut and pinching the bar of the chain saw and to help initiate the fall.

Figure 14.13 The hinge is critical in controlling the fall.

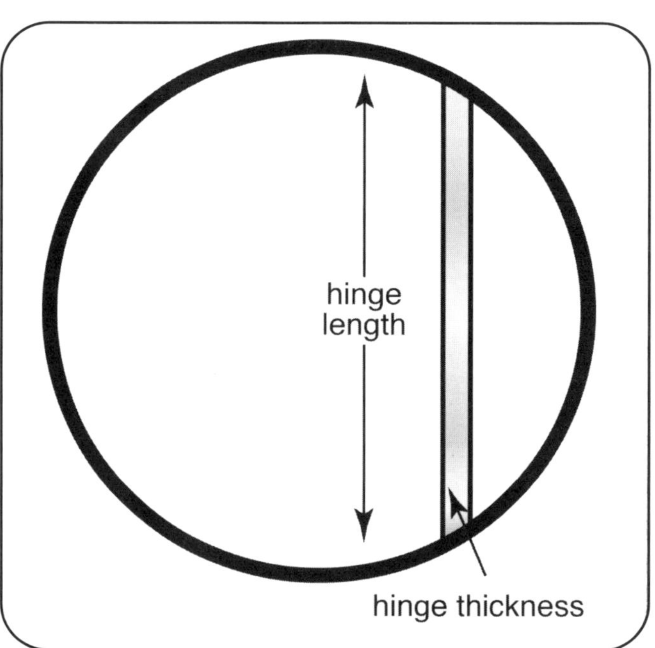

Figure 14.14 The hinge length should be approximately 80 percent of the tree's diameter. The thickness will vary, but approximately 10 percent of the tree's diameter is the rule of thumb.

CHIPPER SAFETY

Brush chippers can be very dangerous machinery, so proper training and safe work practices are essential when operating them. Training should include instruction on daily inspection and maintenance, towing procedures, starting the chipper, feeding brush, and the potential safety hazards involved with operation. Proper PPE is required, and loose clothing, jewelry, climbing saddles, harnesses or body belts, and gauntlet-type gloves should not be worn while operating chippers because they could be caught on brush and could pull the operator into the chipper.

Brush should be fed from the side, and the worker feeding the brush should move away as the brush is fed. The larger, butt end of each branch should be fed in first. Smaller limbs should be pushed into the chipper with larger limbs. No part of the operator's body should ever reach beyond the back edge of the infeed chute (Figure 14.15). The operator should be careful to avoid placing foreign material such as rocks, wires, or other debris into the chipper because such material could damage the knives or cause projectiles to be thrown from the machine.

Many accidents have occurred when workers attempted to perform maintenance while the chipper disk or drum was still moving. No person should ever work on a chipper unless the engine is turned off, the ignition key removed, and the cutter wheel completely stopped and prevented from moving.

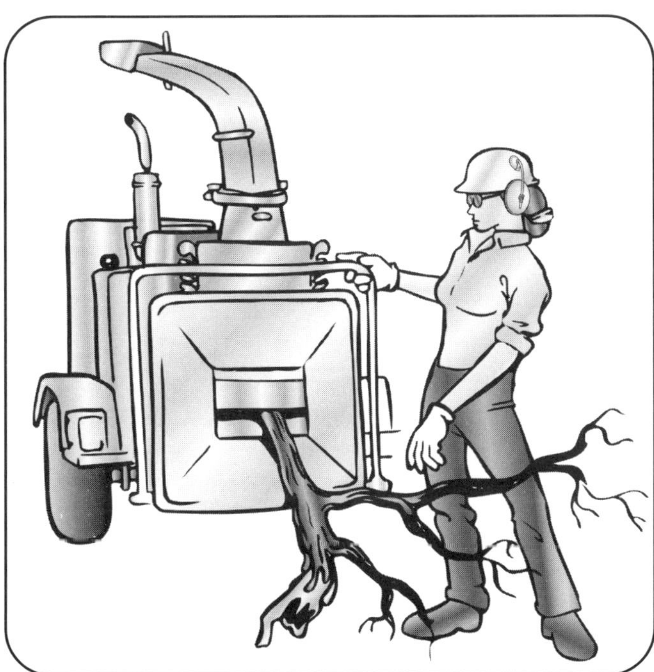

Figure 14.15 Brush should be fed into chippers from the side. The operator must wear the proper PPE.

FIRST-AID PROCEDURES

Urgent Care Directions

1. Rescue the victim from any life-threatening situation.
2. If the victim is not breathing, begin artificial respiration.
3. Control severe bleeding.
4. Do not move the victim unless necessary.
5. If the victim has been poisoned, call a doctor or Poison Control Center immediately.
6. Support fractures.
7. Treat for shock.
8. Call or send for medical help.

Artificial Respiration

1. Listen and feel for breathing by placing your ear and cheek close to the victim's mouth.
2. Check to see that the air passage is clear; remove obstructions if necessary.
3. Tilt the victim's head back with one hand on the forehead. Use your other hand to support the victim's neck.
4. Lightly pinch shut the victim's nose.
5. Take a deep breath. Create a tight seal around the victim's mouth with your mouth. Blow air into the victim's mouth. Give two quick, full breaths. Watch for the victim's chest to rise.
6. Check again for breathing.
7. If victim is unable to breathe without assistance, provide a full breath of air every 5 seconds.

Cardiopulmonary Resuscitation (CPR)

Note: CPR requires special training and should be carried out only by qualified persons. (This section is intended to serve as a reminder for trained persons.)

1. Attempt to rouse the victim.
2. Open the airway by tilting the victim's head back.
3. Look, listen, and feel for breathing. If there is no breathing , . . .
4. Apply artificial respiration (Figures 14.16 and 14.17).
5. Feel for the carotid pulse in the groove of the neck beside the Adam's apple (Figure 14.18). If the pulse is absent, . . .
6. Place victim on his or her back on a flat, firm, horizontal surface.
7. Locate the xiphoid tip (the V in the center of the chest where the ribs meet). Measure two finger-widths up on the breastbone (Figure 14.19).
8. Place one hand over the other. Exert pressure vertically and depress the chest 1½ to 2 inches

with the heel of your hand, keeping your arms and elbows straight (Figure 14.20).

9. After each compression, release pressure completely, but do not remove your hands from that position. Use a 15:2 ratio: 15 compressions followed by 2 quick lung inflations, 80 compressions per minute. Be sure to reposition your hands before resuming compressions.
10. Always continue CPR procedures until the victim is revived or until professional help arrives.

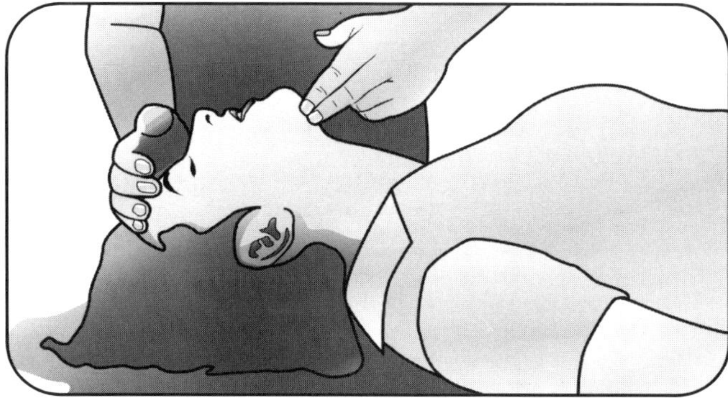

Figure 14.16 Tilt the head back.

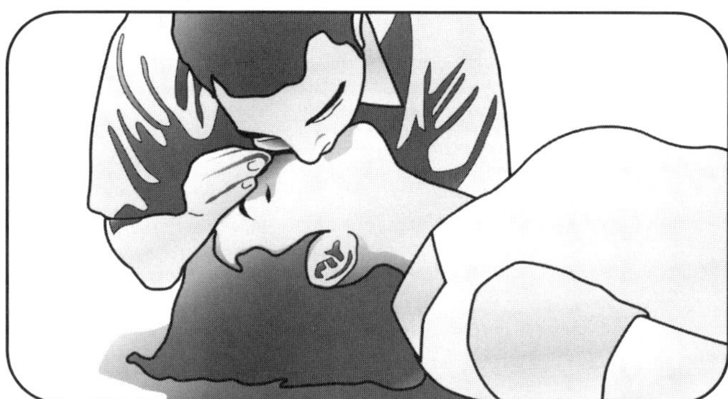

Figure 14.17 Artificial respiration.

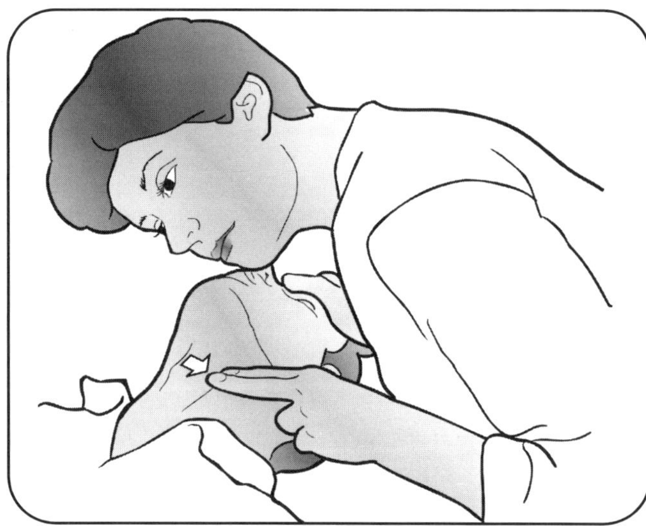

Figure 14.18 Feel for the carotid pulse.

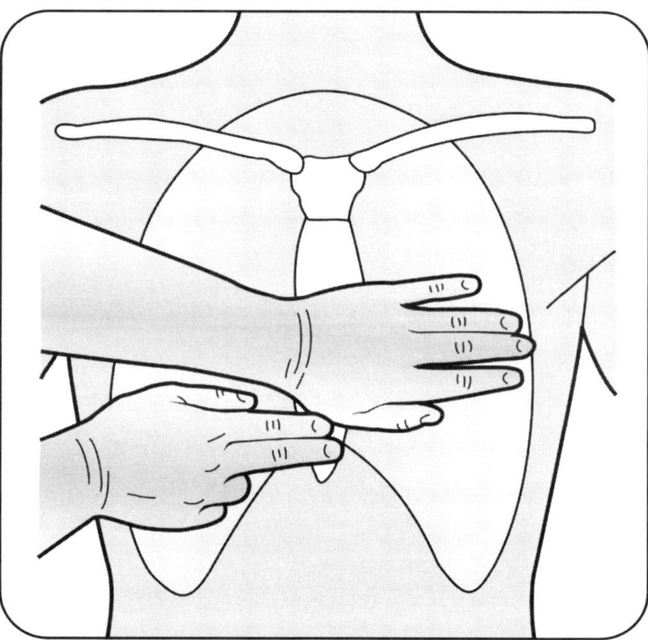

Figure 14.19 Locate the xiphoid tip. Measure two finger-widths up on the breastbone.

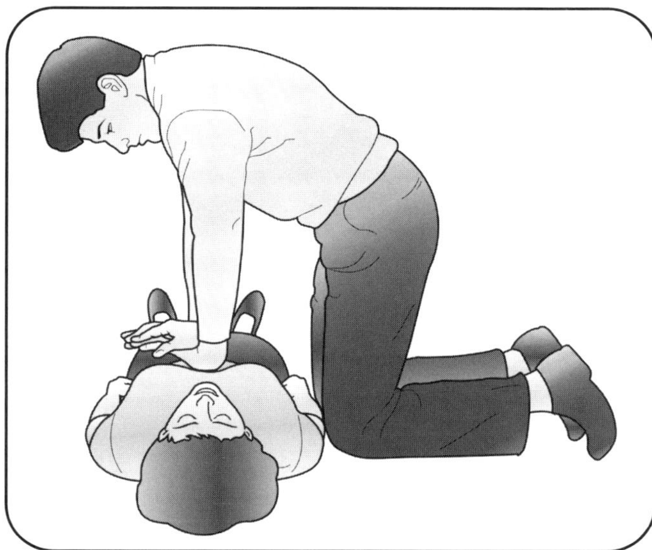

Figure 14.20 Keep arms straight while depressing the chest 1½ to 2 inches.

Severe Bleeding

1. Apply direct pressure firmly on the wound. A thick pad of cloth (sterile, if possible) between your hand and the wound will help control bleeding (Figure 14.21).
2. Elevate the wound if practical.
3. If bleeding continues, apply direct pressure on a pressure point to help stop bleeding from a wound in an arm or leg. Press the main supply artery against the underlying bone.
4. Do not apply a tourniquet unless medical help is available.

Head Injuries

1. Minimize movement and immobilize the victim's neck.
2. Do not attempt to cleanse scalp wounds because this may cause severe bleeding or severe contamination.
3. Place a sterile dressing on the wound, but do not apply excessive pressure.
4. Do not give the victim fluids by mouth.
5. If the victim loses consciousness, get medical help immediately and record the extent and duration of unconsciousness.

Burns

- **Minor burns: first degree (redness, swelling)**
  - Apply cold water or ice.
- **More serious burns: second degree (redness, mottled appearance, blisters, swelling)**
  - Wrap in clean, dry cloth.
  - Do not pop blisters.
  - Do not apply ointment.
  - Seek medical help.
- **Severe burns: third degree (tissue destruction, glossy white appearance)**
  - Cover burn with clean, dry cloth.
  - Seek medical help immediately.

Shock

Symptoms

- pale, cold, moist, and clammy skin
- the whole body weak
- pulse weak and rapid

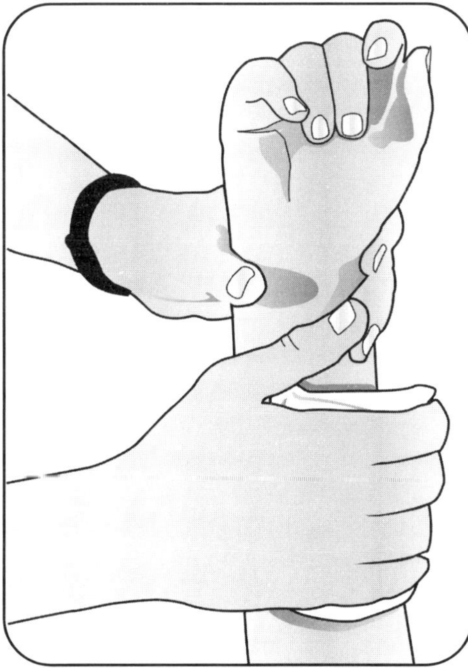

Figure 14.21 Apply direct pressure, preferably with a sterile cloth.

- irregular breathing, often shallow
- nausea, vomiting
- pupils dilated

Treatment

1. Keep the victim lying down. Elevate the feet if this will not aggravate other injuries (Figure 14.22).
2. Keep the victim warm. Cover with a blanket if necessary.
3. Get medical help.

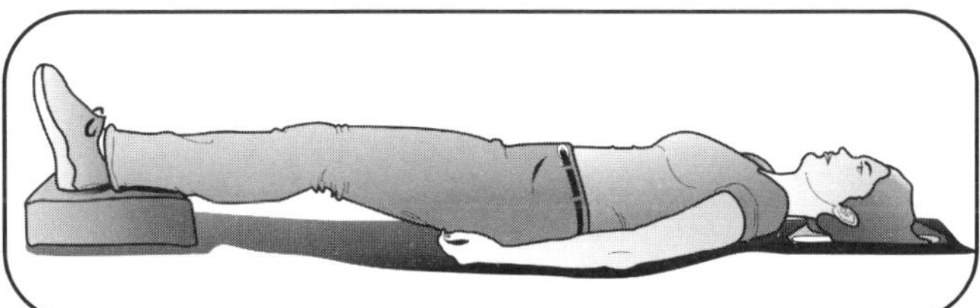

Figure 14.22 Elevate the shock victim's feet if doing so will not aggravate other injuries.

Heat Stroke

Symptoms

- high body temperature
- skin is hot, red, and dry
- rapid pulse
- victim may be unconscious

Treatment

1. Get medical help.
2. Take immediate measures to cool the victim's body.
3. Remove the victim's clothing and cool the body with cool water or rubbing alcohol.
4. Use fans or cold packs if available.
5. Monitor body temperature and reduce it if necessary.

Heat Cramps

1. Firmly massage the cramped muscle.
2. Rehydrate the victim with water or a commercial sports drink.

Heat Exhaustion

Symptoms

- approximately normal body temperature
- pale, clammy skin
- heavy perspiration
- weakness, fatigue
- headache or cramps
- dizziness, nausea, or vomiting

Treatment

1. Rehydrate the victim with water or a commercial sports drink.
2. Keep the victim lying down and raise the feet.
3. Keep the victim cool.
4. If vomiting occurs, discontinue giving fluids.
5. Take the victim to a hospital if symptoms continue.

Poisonous Insect Bites

Minor Bites and Stings

1. Carefully remove the stinger, if present.
2. Apply ice to reduce swelling.
3. Apply soothing lotions.

Tick Bites

1. Cover the tick with heavy oil to close its breathing pores.
2. If the tick does not disengage, leave the oil on for half an hour, then carefully remove the tick with tweezers.
3. Consult a doctor if a rash or other symptoms appear.

Severe Reactions to Bites or Stings

1. Get medical help if the victim is allergic.
2. If the victim has a bee-sting kit, follow directions and use it immediately.
3. Apply a restricting band above the site, if practical. You should be able to slip your finger under the band when it is in place.
4. Keep the affected part of the body below the victim's heart.
5. Apply ice and treat for shock.
6. In the case of a bee sting, remove the stinger with tweezers. Be careful not to squeeze the attached venom sac because this would inject more venom.

Contact with Poisonous Plants

Identification

- **Poison sumac (*Toxicodendron vernix*)**
  - woody shrub or small tree, 5 to 25 feet tall
  - compound leaves, 7 to 11 leaflets
  - fruit: glossy, pale yellow, pendant when ripe
  - distribution: predominantly east of the Mississippi River; swampy areas
- **Western poison oak (*Toxicodendron diversiloba*)**
  - usually in shrub form, sometimes vine
  - leaves composed of three leaflets
  - fruit: white berries
  - distribution: western North America
- **Poison ivy (*Toxicodendron radicans*)**
  - small shrub or vine
  - leaves composed of three leaflets, orange-red in fall
  - fruit: white berries
  - distribution: most of United States, although different varieties inhabit different regions

Treatment

1. Wash affected area with strong soap and water as soon as possible.
2. Wash with rubbing alcohol.
3. Seek medical help if a severe reaction occurs in the next day or two.

Chapter 14 Workbook

1. As used in the ANSI standards, the term ____________ denotes a mandatory requirement, and the term __________ denotes an advisory recommendation.
2. Which personal safety equipment is required for all tree workers?

   a.

   b.

   c.

   d.

   e.
3. True/False—All communications wires and cables shall be considered to be energized with potentially fatal voltages and shall never be touched directly or indirectly.
4. True/False—OSHA (OHSA, in Canada) regulates industrial safety and health issues.
5. The safety standard for tree care operations in the United States is the _______________.
6. True/False—Head protection need only be worn while there are climbers in the trees.
7. True/False—Eye protection is not required for tree work.
8. True/False—Hearing protection may be in the form of ear plugs or ear-muff–type devices.
9. True/False—Workers must not wear gauntlet-type gloves while chipping brush.
10. If a climber warns, "Stand clear," he or she still must not proceed until hearing the acknowledgment, ____ ________.
11. Each job should begin with a _____ _________, which coordinates the activities of every worker.
12. Do not start or operate equipment within ____ feet of the refueling site.
13. All workers should receive some training in ___________ _________ procedures, including CPR, first aid, and aerial rescue.
14. True/False—Any tree workers who work in proximity to electrical conductors must receive training in electrical hazards.
15. True/False—Electrical conductor is defined as any overhead or underground electrical device, including wires and cables, power lines, and other such facilities.
16. True/False—Rubber footwear and gloves provide absolute protection from electrical hazards.
17. Always engage the _______ _______ before starting a chain saw.
18. True/False—On the ground, both hands are required for chain saw operation, but, in the tree, only one hand need be used.
19. True/False—The chain saw engine must be stopped for refueling.
20. True/False—Kickback can occur when the upper tip of the guide bar contacts an object.
21. True/False—A well trained climber, in good condition, should be able to dodge the kickback of a chain saw.

22. Name three types of notches used in tree felling.
    a.
    b.
    c.
23. The depth of the notch should not exceed ________ of the diameter of the tree.
24. The __________ is critical in controlling the direction of fall.
25. Sometimes, if the tree is leaning in the direction of fall or has internal faults, the tree can split upward from the back cut. This is called a ___________ ________, and it can be very dangerous.
26. Workers should feed brush into chippers from the _______ and not allow any part of the body to cross the plane of the infeed chute.
27. True/False—Safety is the responsibility of all employees from the owner to the ground worker.
28. A condition that often follows accidents in which the victim may experience a weak and rapid pulse, shallow breathing, and cool, clammy skin is known as ______________.
29. If a victim is not breathing, it is necessary to perform __________________ ____________________.
30. Three plants that can cause severe skin rashes are ___________ _____, _____________ _____, and ___________ ________.

MATCHING

___ shall	A. leg protection for chain saw use
___ approved	B. advisory recommendation
___ CPR	C. cardiopulmonary resuscitation
___ direct contact	D. body touches energized conductor
___ should	E. safety standards for tree work
___ indirect contact	F. mandatory requirement
___ ANSI Z133.1	G. meets applicable standards
___ chaps	H. touching an object in contact with an energized conductor

CHALLENGE QUESTIONS

1. Explain how the ANSI Z133.1 safety standards are developed and how members of the tree care industry can have input.

2. Diagram three tree-felling notches and appropriate back cuts for each. Explain the role of the hinge in tree felling.

3. What are some of the most common causes of injuries in tree care, and how can they be prevented?

SAMPLE TEST QUESTIONS

1. According to ANSI Z133.1 standards, the term *shall* denotes
   a. an advisory recommendation
   b. a mandatory requirement
   c. a safety suggestion by NAA
   d. none of the above
2. Which of the following should be considered energized with a potentially fatal voltage?
   a. overhead electric lines
   b. underground electric lines
   c. telephone and cable TV wires
   d. all of the above
3. Head protection is required for tree workers
   a. whenever performing tree care operations
   b. when specified by the supervisor
   c. whenever there are climbers working aloft
   d. only if chain saws or chippers are in use
4. The situation that can cause chain saw kickback is
   a. failure to maintain adequate chain tension
   b. a worn sprocket or guide bar
   c. the upper quadrant of the guide bar contacts an object
   d. uneven sharpening of the chain saw teeth
5. Which of the following should be treated first?
   a. severe bleeding
   b. suspended breathing
   c. poisoning
   d. heat stroke

Other Sources of Information

(See pages v–vi for complete bibliographic information.)

American Red Cross, 1992. *First Aid & Safety Handbook.*
ANSI Z133.1. *Safety Requirements for Pruning, Trimming, Repairing, Maintaining, and Removing Trees and for Cutting Brush.*
ISA. *ArborMaster Video Series III.*
NAA/ISA, 2000. *Basic Training for Ground Operations in Tree Care.*

CHAPTER 15

CLIMBING AND WORKING IN TREES

CHAPTER 15 CLIMBING AND WORKING IN TREES

CHAPTER 15

CLIMBING AND WORKING IN TREES

Objectives

1. Become familiar with the American National Standards Institute's ANSI Z133.1 standards for tree care operations.
2. Become familiar with the safety equipment used for climbing and working in trees. Learn the ropes, knots, and tools used by tree workers.
3. Learn the techniques used by tree climbers and be able to recognize safety hazards.
4. Discuss the principles of rigging and the equipment and techniques involved.

Key Terms

3-strand
12-strand
16-strand
ANSI Z133.1
access line
aerial rescue
arborist block
back cut
balance
bend
bend ratio
bight
Blake's hitch
block
body thrust
bollard
butt-hitching
butt-tying
carabiner
climbing hitch
climbing line
climbing saddle
climbing spurs
clove hitch
cycles to failure
design factor
double braid
double crotch
drop cut
drop zone
dynamic loading
false crotch
figure-8 knot
footlocking
friction device
half hitch
hinge
hinge cut
hitch
hollow braid
kerf
kernmantle rope
landing zone
lanyard
load line
lowering device
micropulley
notch
personal protective equipment (PPE)
pole pruner
pole saw
positive-locking
Prusik hitch
Prusik loop
redirect rigging
rescue kit
rescue pulley
rigging
rigging line
rigging point
rope sling
running bowline
scabbard
screw link
secured footlock
shackle
sheave
shock-loading
snap
snap cut
speedline
split-tail
standing part
stopper knot
tagline
tautline hitch
tensile strength
throwing knot
throwline
tip-tied
webbing sling
working end
working load limit (WLL)
work-positioning lanyard

INTRODUCTION

Tree climbing is a very physical and potentially hazardous profession. However, a well-trained worker who follows all established safety standards and procedures can work safely and efficiently in a tree. Before climbing a tree, a climber should first inspect all safety equipment. Then the tree itself should be inspected for hazards. A good climber plans ahead where to tie in and how to work the tree. A little forethought saves energy and may prevent accidents. Tree climbers should be familiar with and comply with all applicable safety standards, particularly the ANSI Z133.1.

INSPECTION OF GEAR

A tree climber's safety depends upon the reliability of the safety gear. A worker's safety gear is called **personal protective equipment (PPE)**. This equipment includes a hard hat, safety glasses or goggles, hearing protection, and personal protective clothing. Approved head and eye protection is to be worn at all times by workers engaged in tree operations.

All equipment used by tree workers, including climbing gear and tools, must conform to applicable safety standards and should not be altered. Equipment should be inspected according to manufacturers' guidelines. **Climbing saddles** should be checked for excessive wear and to see that stitching and rivets are strong and intact.

Snaps used in securing the **climbing line** or **lanyard** must be self-closing and self-locking. **Carabiners** used for climbing must be self-closing and **positive-locking**. Both must have a minimum **tensile strength** of 5,000 pounds. Snaps and carabiners should be checked before and during use to see that they are functioning properly.

Climbing lines must be identified by the manufacturer as suitable for tree climbing with adequate strength, wear, and stretch characteristics. With some exceptions, climbing lines must be ½-inch diameter, constructed of synthetic materials, and have a minimum tensile strength of at least 5,400 pounds. Climbing lines should be inspected before each use. Check for cuts, abrasions, changes in diameter, discoloration, or glazing of the fibers. If a snap is used, it should be routinely moved to the opposite end of the line. If one end of the climbing line shows signs of excessive wear, it should be cut off. Old, worn, or cut ropes must be retired from use. **Work-positioning lanyards** must also be inspected carefully before each climb. They must meet strength requirements for ropes and snaps. Look for abrasions, excessive wear, or faulty snaps. **Prusik loops** and **split-tails** used in a climbing system must meet the minimum strength standards for climbing lines (Figure 15.1).

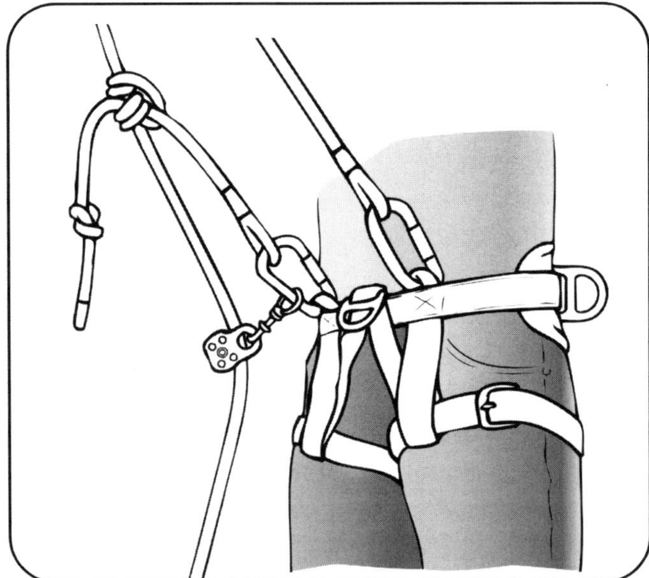

Figure 15.1 This drawing illustrates dual tie-in points using double-locking carabiners, a split-tail for tying in, and a micropulley fair lead for tending the climbing hitch.

All equipment used by tree workers should conform to the applicable ANSI (American National Standards Institute) standards for tree workers, or equivalent standards within their jurisdiction (OSHA or CSA). All equipment should be inspected according to applicable ANSI guidelines and manufacturers' recommendations.

KNOTS

All tree workers should be familiar with the knots used in tree work. A climber should know how to tie and untie each of these knots. Part of knowing how to tie a knot is knowing how to "dress" and "set" the knot properly. The dressing of the knot is the aligning of the parts; setting it tightens the knot in place. A climber must know how each of the common knots is used, and the advantages and disadvantages of each.

"Knot" is the general term given for all knots, hitches, and bends. A **hitch** is a type of knot used to secure a rope to an object, another rope, or the standing part of the same rope. A **bend** joins two rope ends together. There are several categories of knots, hitches, and bends. Tree climbers use end-line knots, hitches, and bends to secure the climbing line to carabiners or rope snaps. End-line knots are also used to tie off branches being rigged.

A type of knot important in tree climbing is the **climbing hitch**. Climbing hitches are the "climbing knots" used by climbers to tie in. For years, the primary climbing hitch used by climbers in the United States has been the **tautline hitch**. **Blake's hitch** is growing in popularity because it maintains more uniform friction and does not roll out. Many European climbers use variations of the **Prusik hitch** to secure themselves when climbing. The knots commonly used in tree care are illustrated in this chapter, along with a brief description of each (Figures 15.2 through 15.12).

INSPECTION OF THE TREE

Every job should begin with a job briefing that covers the work plan, potential hazards, and all required gear and procedures. Before climbing a tree, a climber should always look carefully and locate any electrical conductors or utility lines. Check for hazards such as dead or broken limbs, cracks, weak branch unions, or signs of decay such as conks or fruiting bodies. Always check the root crown of a tree as well. Soil, bark, or ivy may hide signs or

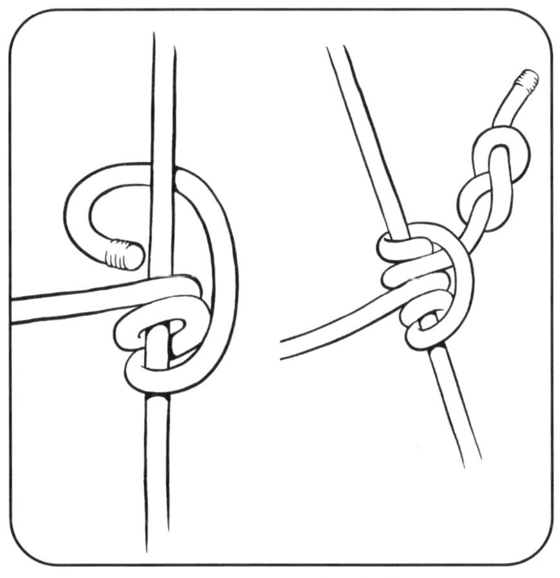

Figure 15.2 Tautline hitch.

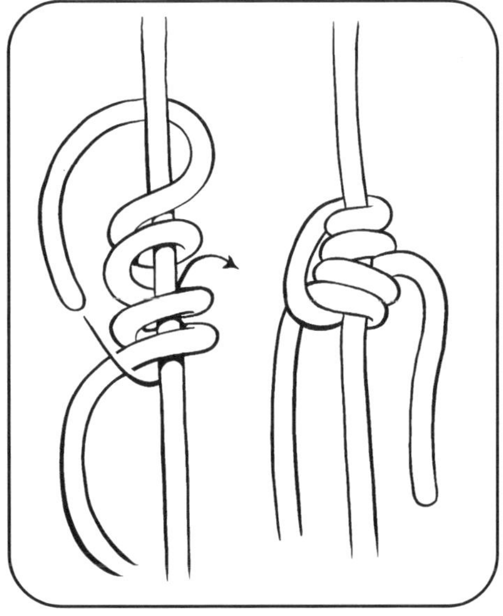

Figure 15.3 Blake's hitch.

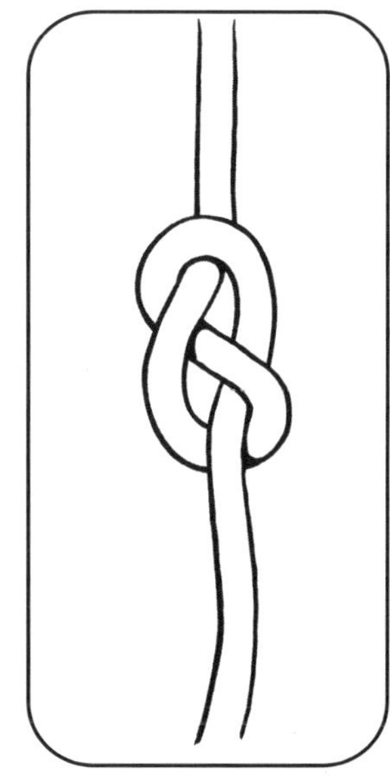

Figure 15.4 Figure-8.

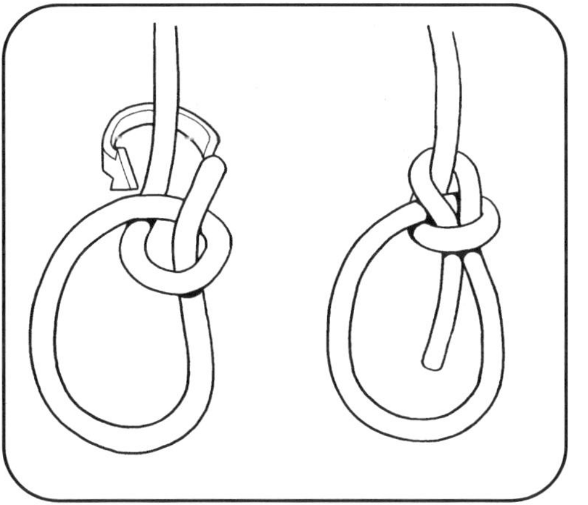

Figure 15.5 Bowline.

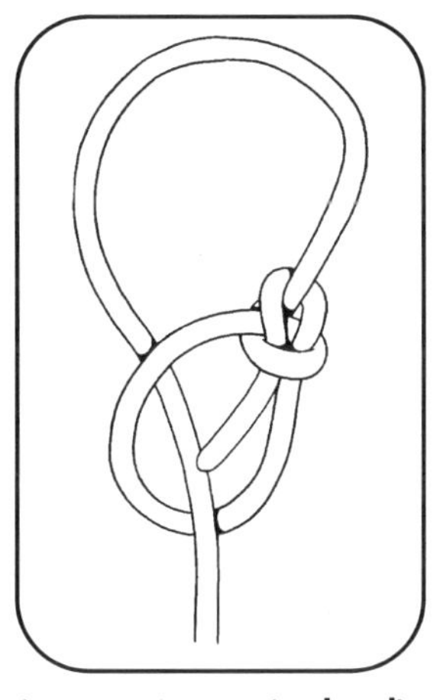

Figure 15.6 Running bowline.

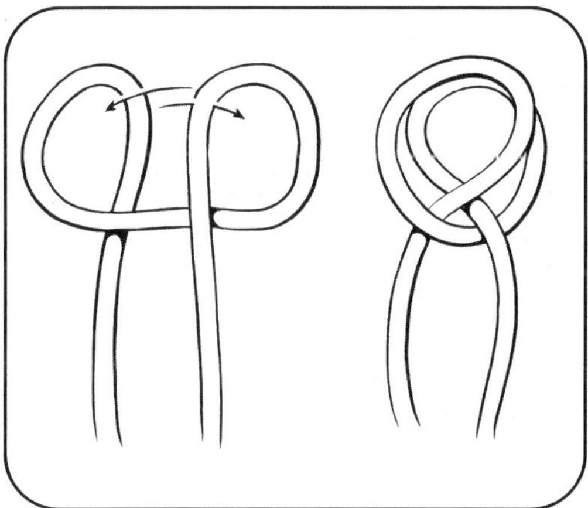

Figure 15.7 Midline clove hitch.

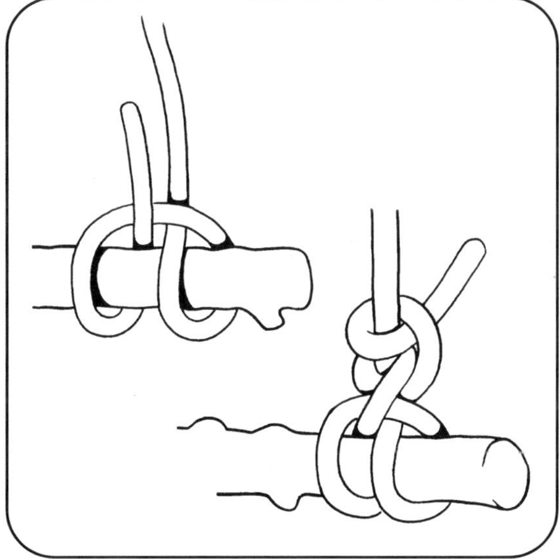

Figure 15.8 End-line clove hitch and two half-hitches.

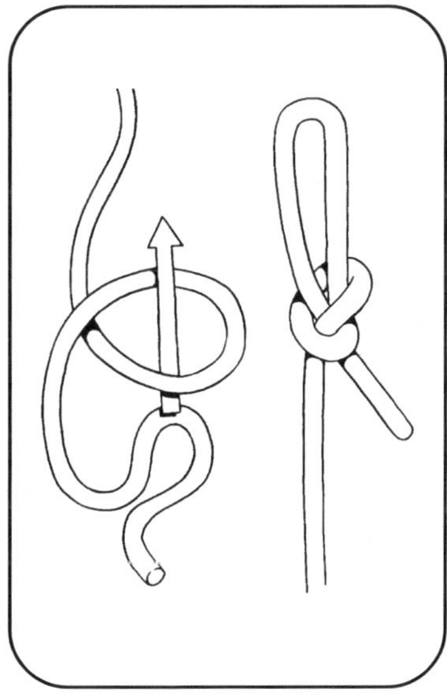

Figure 15.9 Slip knot.

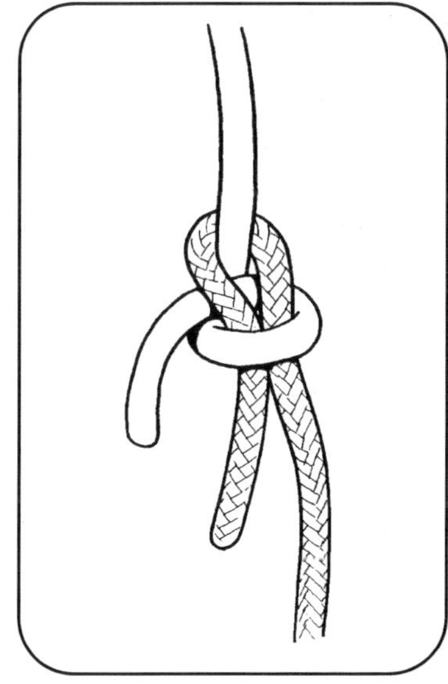

Figure 15.10 Sheet bend.

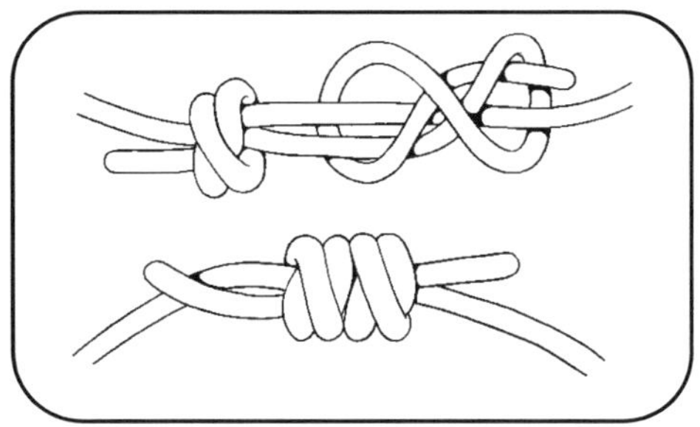

Figure 15.11 Double fisherman's knot.

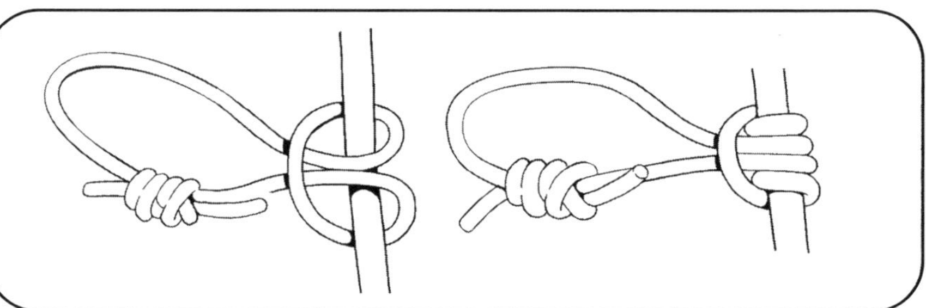

Figure 15.12 Prusik hitch. Note: A three-wrap, six-coil Prusik often is used with a Prusik loop.

symptoms of decay. Severe crown rot may result in the tree falling over. (Chapter 12 of this text provides more detailed information about potential defects.)

The pre-climb inspection should also be used to plan how the tree will be climbed. It is usually a good idea to plan the climbing route while still on the ground and to choose a safe tie-in point from which the tree can be accessed. An experienced climber will also become familiar with the characteristics of various trees. It is essential to know how strong or brittle a tree's wood is.

CLIMBING TECHNIQUES AND PROCEDURES

Once the tree, site, and climbing gear have been inspected and deemed to be safe, the climber can plan a climbing strategy. There are many ways of getting into and ascending a tree. The climber can use the climbing line, a ladder, or **climbing spurs** (if the tree is to be removed). Each method has advantages and limitations. With few exceptions, a climber must be tied in or otherwise secured while entering or working in a tree. One technique is to use two lanyards (or a 2-in-1 lanyard) so that the second can be secured when the first must be moved around a limb. The climber should climb on the side of the tree away from electrical hazards, if present.

To set a rope in the tree, a climber may choose to use a **throwline** (Figure 15.13). A shot pouch attached to the throwline can be thrown with amazing accuracy through crotches 60 feet or higher. Shot pouches are available in various weights, and selection is mostly a matter of preference. Arborists have developed a number of techniques for throwing, launching, and manipulating throwlines to put them in the specific tree crotch desired. After the throwline passes through the crotch, it falls to the ground. Sometimes the climber must manipulate the cord to encourage the weight to come down. The climber's line can then be attached to the cord of the throwing weight and pulled through the crotch.

Figure 15.13 A throwline can be used to set a line in a tree.

Sometimes a climber may throw the climbing line directly into the tree. On short, open throws, it may be easiest to simply loop the rope over a low limb. For trickier throws, the climber may use a **throwing knot.** A throwing knot is simply a series of wraps that hold the rope together and provide end weight to facilitate throwing. The throwing knot can be used in an open or closed form; the closed version will not come undone when the rope is thrown.

Once a rope has been set in the tree, there are several methods of ascending. One method is the **body-thrust** technique, in which the climber uses the rope to climb the tree. Another method is the **secured footlock** technique, in which the climber climbs the rope itself. Other alternatives are variations of the two, or techniques that employ mechanical ascending devices.

In body thrusting, the climber attaches one end of the climbing line to the center D-ring(s) of the saddle. This can be accomplished using a locking snap, double-locking carabiner, or a variety of knots. After attaching the rope to the hardware, a tail piece of rope is left (or a separate piece is used) to form the climbing hitch around the standing part of the climbing line. Technique is important in the body thrust. The climber should place his or her feet high on the trunk of the tree. The hips are thrust upward, creating slack in the line, and simultaneously the other side of the line is pulled down, taking up the slack and keeping the line taut (Figure 15.14). The climber who relies solely on upper body strength to body thrust can be exhausted after reaching the top. The addition of a **micropulley** below the climbing hitch allows a ground worker to advance the knot and pull slack out of the climber's line while the climber ascends (Figure 15.15).

Footlocking is another method of ascending a tree once a rope has been set. When footlocking, the climber actually climbs the rope and might not contact the tree until the top. If the footlocking method is used, the climber should use the secured footlock technique. In the secured footlock technique, the use of a Prusik loop makes footlocking much safer. The Prusik loop is tied to the climber's line using a Prusik hitch (or a mechanical device) and attached to the front D-rings of the saddle, using an approved double-locking carabiner or snap. This serves as a means of securing the climber.

Standing with hands high on the climbing line and with the Prusik hitch above the hands, the climber grasps the rope, raises one foot, and aligns the rope on the inside of his or her knee and across the top of the instep. Then, with knees apart, the feet are raised high and the second foot pulls the rope from below and over the first foot. The rope is locked off by standing on top of the section of rope that is wrapped around the foot. The climber then stands up, grabbing with the hands, and the process is repeated (Figures 15.16 and 15.17). Once at the top of the rope, the climber must transfer into the tree. This is a potentially dangerous transfer, so the climber should either tie in or use a work-positioning lanyard before removing the Prusik. If the footlocking rope has been set above a lower limb, the climber will be able to enter the tree onto the lower limb, and transfer will be facilitated.

Another method of ascending a tree is the use of climbing spurs. Because spurs can damage a tree, they are approved for use only on trees to be removed, or for aerial rescues. A climber should never spur up a tree without using a work-positioning lanyard, and/or being tied in. Lanyards are available in a number of lengths and styles. Some have a wire core that makes the line more rigid. This type should never be used around electrical hazards. It should also not be assumed to provide protection from cutting with a chain saw.

Figure 15.14 The body-thrust technique.

Figure 15.15 Adding a micropulley allows the ground worker to advance the climbing hitch while pulling slack.

When ascending large trees, it may be necessary to reset the climbing line several times. Whenever

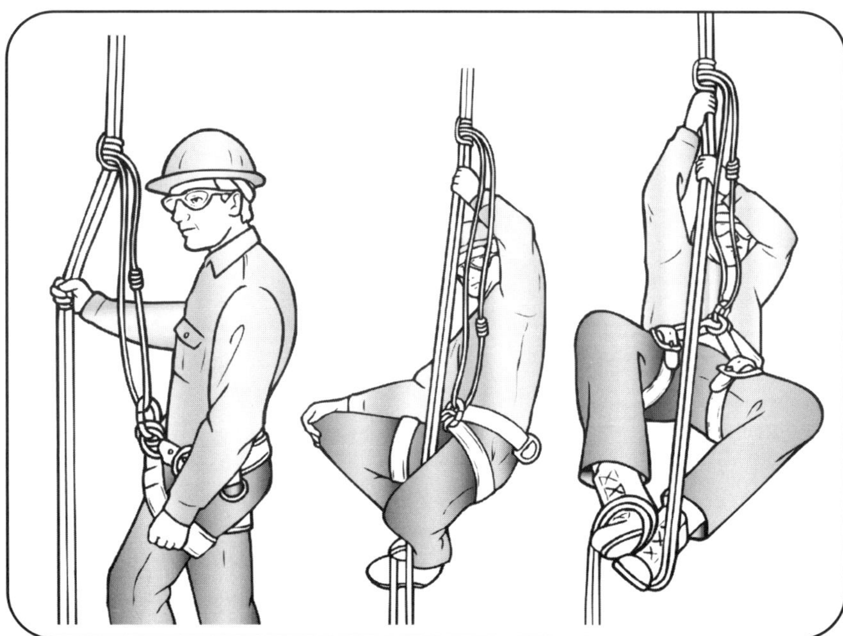

Figure 15.16 The secured footlock technique.

Figure 15.17 A climber can footlock the running part of the line while tied in but away from the trunk.

untying, the climber should be secured with a work-positioning lanyard. Another technique is to alternate the ends of the climbing line when recrotching. This way, the climber is still tied-in while recrotching the other end.

Often the climbing line must be set in a crotch well above the climber's head. One technique is to throw the rope over a higher limb. Another method is to set the rope higher in the tree using a pole. When using a pole saw or pole pruner, be sure to keep the rope on the back side, away from the cutting portion. Some climbers use a pole saw with no blade, just a hook, for setting ropes.

Tying-In

The choice of where to tie in is very important. Generally, it is desirable to pick a high, central location in the tree. This allows freedom of movement and easy access to most points below in the tree. The higher the tie-in point, the farther the climber can move out on the limbs (Figure 15.18). It is easiest to work when tied in directly above the working area. The more vertical the climbing line, the more secure the climber. It is very important not to tie in to a crotch that would allow a swing toward power lines in the event of a slip or fall.

The crotch selected for tying in should be wide enough for the rope to pass through easily. The size of the limbs vary with species and wood strength, but generally the main branch should be at least 4 inches in diameter. The climbing line is tied in by passing it through a crotch, around the larger limb or trunk, and over the smaller or lateral branch. This way, if the smaller branch breaks clean, the rope will simply drop to the next branch down, rather than out of the tree. Climbers may also choose to use a **false crotch** when tying in. This can reduce the wear on the rope and damage to the tree and can, in some cases, facilitate climbing.

Figure 15.18 Tying-in at a point that is high and central in the tree allows the climber to work on most limbs.

The climber ties in by tying a climbing hitch from the tail of the attachment to the D-rings of the climbing saddle, to the other side of the climbing line. A **figure-8 knot** should be tied in the tail from the climbing hitch as a **stopper knot** to prevent the end from going through the climbing hitch. It is a good idea to use a climbing line that is long enough to allow the climber to reach the ground. If there is any question of this, the climber should tie a stopper knot in the opposite end of the climbing line when working in tall trees to prevent the end from passing through the climbing hitch if the climber reaches the end of the line.

Sometimes it is helpful for a climber to **double crotch**. Double crotching is simply tying in at a second crotch using the far end of the climbing line, or a second line (Figure 15.19). The primary use of this technique is in ascending a tree. The climber can be secured against a fall while crotching the climbing line at a higher point. The lower tie-in would be untied as the worker climbs past. The double-crotching technique may also be used if the climber is ascending a second leader in a tree with a wide spread. The climbing line can be used to help climb the upright limb without sacrificing the original tie-in point. Another use of the double-crotching technique is to allow the climber to be suspended between limbs. This can be useful for installing cables, working on hazardous lower limbs, working on storm-damaged trees, or transferring from one tree to another.

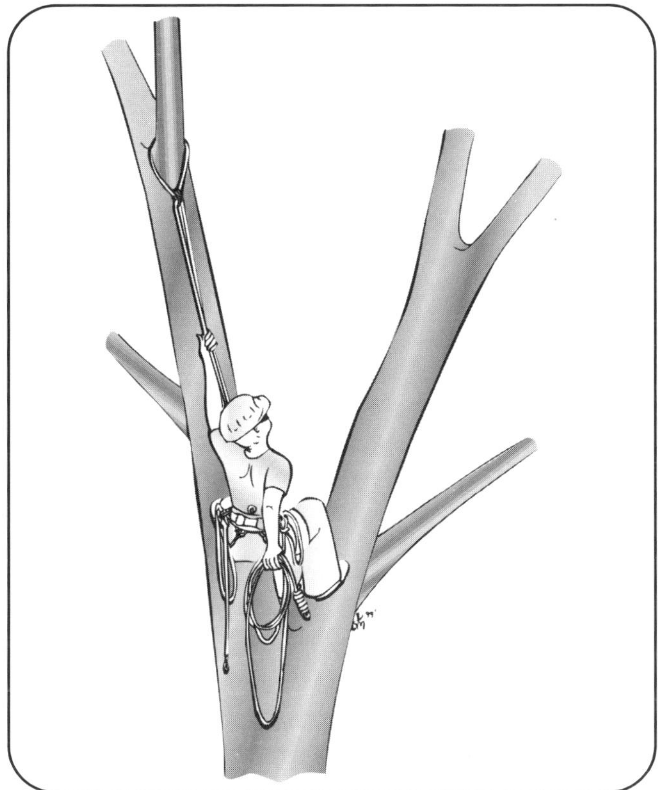

Figure 15.19 The climber is tied in using a friction-saving false crotch on the left limb, and is preparing to double crotch on the right limb.

Use of the Climbing Line and Work Positioning

The climbing line is more than a safety device to secure the climber against falling. A good climber uses the rope to ascend the tree, access branch tips, maintain balance, and move freely within the tree. In addition, the rope enables the climber to use both hands, without any loss of stability (Figure 15.20). If the climber is able to keep his or her weight on the climbing line, both hands can be used for working. To be stable, the three-point contact rule should be followed. The climbing line, when taut, can be considered one point of contact. The lanyard, if used, is also a contact point. Whenever the climber's weight is not on the climbing line, a three-point contact should be maintained with the tree.

Figure 15.20 Once tied in, the climber's hands are free for working.

A climber can walk out on limbs for access to tips for proper thinning. Generally, the preferred method is to walk backward or sideways on the limb, keeping tension on the line (Figure 15.21). Whenever the climber is out toward the tip of a horizontal limb, it is important to keep his or her weight on the rope. If the climber allows his or her weight to be on the limb, the limb may break. The angle of tie-in is important. As a rule, the higher above the work station the tie-in point is, the greater distance the climber can move out from the trunk (Figure 15.22). Another technique in which the climber relies on the rope is swinging. When suspended on his or her climbing line, the climber can sometimes swing like a pendulum to reach other

Figure 15.21 Limb walking. It is important to keep tension in the climbing line for balance.

Figure 15.22 Keeping weight on the climbing rope allows the climber to go out near the ends of limbs.

parts of the tree. Control is crucial when swinging, to avoid crashing back into the trunk of the tree.

WORKING IN A TREE

While in a tree, the climber may require various tools and equipment, including a chain saw, **pole pruner**, **pole saw**, cabling hardware, and/or other tools. Most climbers climb with their handsaw and **scabbard** (sheath for the handsaw). The ground workers send up other tools. Workers may tie equipment onto the climber's line using a **clove hitch**. If pole pruners or pole saws are used in the tree, they should be hung vertically in the tree when not in use. They should be hung in such a way that the sharp edge is away from the worker, and so that they will not accidentally dislodge.

Chain saws can be equipped with a chain saw lanyard for use in a tree. Chain saws weighing more than 15 pounds should be supported by a separate line when used in a tree. The climber must be stable and secure when using chain saws and other equipment in the tree. Chain saws should be shut off when the climber moves to another position. Because it is extremely dangerous to use a chain saw in a tree, safety precautions are important. A climber should be secured with a work-positioning lanyard in addition to the climbing line when using a chain saw in a tree. This is for added stability and safety in case the climbing line is severed accidentally.

When pruning a tree, the climber usually works from the top down. Limbs may be pruned evenly out to the tips. The climber normally works radially around the tree to access each limb, while working down through the tree. Normally, the lowest limbs are the last to be pruned, but this may vary from tree to tree. The climber should think ahead and plan the descent, especially in large trees.

EMERGENCY RESPONSE AND AERIAL RESCUE

A tree climber must take many precautions to guard against accidents. Accidents are prevented through the conscious recognition of potential hazards in the workplace, and the effort to avoid them. Yet it takes only one lax moment or unexpected event for an accident to happen. Because of this, every worker on the crew should be trained in first aid, cardiopulmonary resuscitation (CPR), and **aerial rescue**. Aerial rescue is the process of safely bringing an injured or unconscious worker to the ground.

The most important aspect of aerial rescue is safety. Workers must be trained to assess the emergency situation and make decisions based on the circumstances, the victim's perceived condition, and the help that may be available. Good training and practice help workers handle emergencies more safely and efficiently. There is no time for panic. A rescuer who fails to take the proper precautions may become a second victim.

There are a number of ways a climber can be injured in a tree. Electrocution, heart attack, heat exhaustion, insect or animal attack, a blow from a swinging limb, or a chain-saw cut could leave a worker dangling helplessly in a tree. Ground workers should maintain a close watch on climbers. A climber could get hurt and lose consciousness without ever calling for help.

When a climber is injured or unconscious in a tree, the rescue procedure should begin immediately.

If there is more than one worker in the area, one should go for help immediately. Emergency numbers should be posted on the dashboard of the vehicle. When calling for emergency assistance, be sure to give the exact location of the accident and the nature of the emergency. Do not hang up first. Let the emergency personnel obtain all necessary information and be the first to hang up. If there is only one rescuer, he or she may call for assistance but should stay and help the injured worker if possible.

The first step in assessing the emergency situation is to determine whether there is an electrical hazard (Figure 15.23). Because the chance of the rescuer becoming a second victim is great, utility company experts recommend that the local electric company be called to avoid any further direct or indirect contact. If the victim has been electrocuted, the rescuer must make an informed decision whether to attempt a rescue or wait for the utility's emergency help. Minutes can mean the difference between life and death. Yet a hasty rescue attempt may lead to the electrocution of the rescuer. Never attempt to climb a tree or rope that may be energized.

Figure 15.23 Before performing an aerial rescue, the rescuer must first determine whether an electrical hazard exists.

If there is no electrical hazard, it is important to get to the victim to assess his or her condition. The rescuer should use proper climbing equipment and remain secured while climbing to the victim. When practical, the rescuer should use a second climbing line and tie in above the victim. If the tree is not energized, climbing spikes may be used to reach the victim.

Upon reaching the victim, a quick check should be made to determine the nature of the injury (Figure 15.24). If the victim appears to have a broken neck or spinal injury, no attempt should be made to move the victim. Be sure the victim is secure, and get emergency help immediately. One of the first tenets of first aid is to avoid moving the victim unless necessary. Although all first-aid procedures can be performed more effectively on the ground, moving the victim may complicate injuries. However, an injured worker hanging for a prolonged period in the climbing harness could lose consciousness and/or go into shock. The rescuer must exercise good judgment based on training and the severity of the situation when deciding whether it is necessary to move the victim. In some cases, the best decision may be to await emergency personnel who will have equipment that can prevent further injuries while lowering the victim. Most emergency rescue teams are not trained or equipped to rescue victims out of trees, however. In many cases, it will still be up to the tree workers to get the victim down.

Figure 15.24 Upon reaching the victim, the rescuer should determine the nature of the injury and first-aid needs.

If the victim is not breathing, artificial respiration should be initiated immediately. Clear the victim's airway, pinch the nose closed, and give several quick breaths through the mouth. If there is no pulse, get the victim to the ground as quickly as possible. CPR cannot be performed in the tree. If there is severe bleeding, elevate the wound and apply direct pressure. If the serious bleeding is from a head wound, do not apply pressure. Talk to the victim while performing the rescue. Be as reassuring as possible.

If it is necessary to bring an injured worker down, it is best for the rescuer to crotch a second climbing line above the victim. It must be in a location that is strong enough to support the weight of the climber and the victim. If the victim's rope is damaged, the rescuer can bring him or her down on the rescuer's line. Otherwise, the climber can operate both climbing lines if necessary. The rescuer should check the victim's rope and saddle for damage, and secure the victim before attempting to lower him or her. Often the best method for bringing the victim down is for the rescuer to attach the victim's D-rings to the rescuer's own and cradle the victim across his or her lap.

The rescuer should support the victim while descending. Attempting to come down too fast may result in further injury to the victim. If practical, a ground worker can lower the injured climber with the climbing line. The ground worker should take a wrap with the end of the rope around a tree or other device for friction to ensure smooth lowering. Then the rescuer in the tree can untie the victim's climbing hitch.

Each member of the crew should be thoroughly trained in first aid and aerial rescue. The necessary rescue equipment must be in good condition and readily available. Some companies keep a separate **rescue kit** that is not used for routine, daily work. This should include a climbing line and saddle, a lanyard, a throwline, climbing spurs, a pole pruner, a sharp knife, and a first-aid kit. The rescue kit should be taken off the truck at the start of each job. It may not be accessible if it is on an energized truck.

Some companies now advocate that a second **access line** be hung when working above 50 feet, particularly if the tree is difficult to enter or ascend. This can save valuable minutes if an aerial rescue becomes necessary.

It is not always possible to foresee an accident. The ability to react swiftly and safely to save a life depends on keeping a cool head, using common sense, and being prepared. Proper training and practice can save crucial minutes that could mean the difference between life and death.

RIGGING

Rigging is the use of ropes and other equipment to take down trees or remove limbs. Rigging is necessary when free-falling is not possible due to potential hazards such as obstacles below or power lines. Rigging techniques often allow the climber to remove larger limbs in less time and with more control.

The basic tools and techniques of rigging are described in this chapter. However, it is emphasized that only experienced tree workers should attempt to use these techniques. Rigging is the most advanced aspect of working in a tree. Even for veteran arborists, it is always best to practice new techniques in open areas, so that safety and control will not be in question on those trees where experience counts.

The tools and techniques used in rigging vary with the situation. If removing limbs from a tree that is being pruned, care must be taken not to damage the remaining branches and trunk. If rigging for removal, the climber has more options in how to work the tree. The simplest form of rigging involves the use of ropes, wraps on the trunk, and natural crotches used as **rigging points**. More sophisticated techniques require a better understanding of the advantages and limitations of the equipment and methods involved.

Rope

Rope may be considered the arborist's most important tool. The characteristics of a rope (strength, stretch, durability, etc.) are the result of the materials and techniques used to make it. To date, polyester is the most widely used material by arborists, and most commercially available climbing and rigging lines are made from this fiber.

Many types of ropes are used in arborist rigging. **Three-strand** rope has relatively low strength and high elongation, and is relatively inexpensive. It is appropriate to run through natural crotches, for climbing or rigging, but it also runs well through a false crotch. A major drawback to 3-strand rope is the twisting, or hockling, that occurs as the line is used. **Sixteen-strand**, braided arborist lines have relatively large cover strands for strength and abrasion resistance and a parallel core to keep the rope round and firm under load. In this construction, the core does not carry the load. These ropes are appropriate for natural crotch climbing or rigging.

Double-braid lines are just that: a rope inside a rope. The core and cover are balanced and share the load almost equally. For this reason, they are not recommended for natural crotch rigging, where the friction of the cover with the tree causes an imbalance in the load taken by the core and cover braids. It is an exceptionally strong and low-stretch line but should be run only over smooth sheaves or bollards. **Hollow braid**, such as **12-strand**, is braided rope without a core. The number and diameter of the strands compared to the diameter of the rope determine if the lines are spliceable or not. They also determine abrasion resistance and whether the rope remains round (Figure 15.25). **Kernmantle rope**, literally meaning core and cover, is adopted from recreational rock climbing and has a very

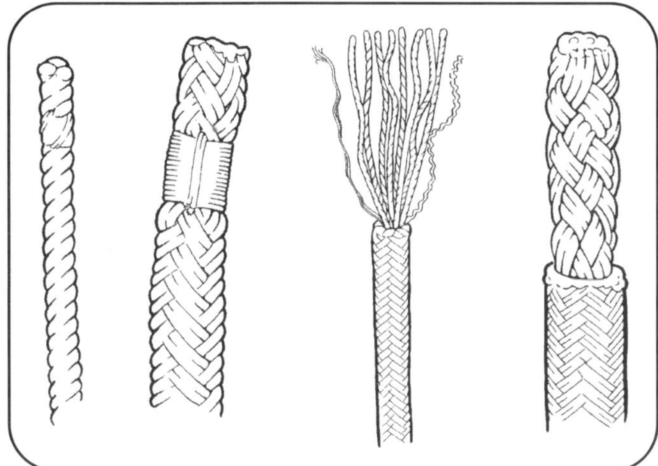

Figure 15.25 Basic rope constructions. From the left are 3-strand, hollow braid, 16-strand with a parallel core, and double braid. The 3-strand and 16-strand are single end, while the others are double end. The core is exposed in the last two.

tightly braided cover to protect the load-bearing core from abrasion. Such ropes are not often used by arborists.

Design and Limitations

Understanding the design and limitations of each piece of equipment employed is vital to the ability to set up a safe and efficient rigging system. Too often, arborists fail to think through the forces and loads that might be involved in rigging out large limbs. The failure of any link in the system can have disastrous effects, including property damage, injuries, or even a fatality. Few practices in arboriculture are as involved or as inherently dangerous as rigging; professional knowledge and experience are essential.

The tensile strength reported by the manufacturer is the breaking strength of a rope or piece of hardware. As a rope is used, its strength is reduced due to dirt, wear, knots, and, of course, loading. **Cycles to failure** must also be considered. One cycle means one lift, or drop, for a rigging line. Each cycle creates permanent damage in the rope, and eventually the rope will fail. At larger loads, this number of cycles to failure is reduced.

If the load in a rope is equal to its tensile strength, the rope may fail when used only once (one cycle to failure). To increase the life of a rope, a **working load limit (WLL)** much less than the tensile strength is established, and workers must ensure that the loads in each cycle are less than this WLL. The **design factor** (sometimes called safety factor) is the tensile strength divided by the WLL. For arborist rigging, with dynamic loading, high wear, and dirty conditions, a design factor of 10 or greater is recommended.

Equipment

Choosing the correct equipment for a given situation can make the job much more productive, and safer, provided the science behind a given device is understood. The demands placed on arborists' equipment by dynamic loading and abrasion mean that tools from other industries are not always applicable to tree care.

Arborists rely on friction to help control loads when lowering branches out of a tree. Historically, **rigging lines** were wrapped around the trunk of the tree to add some friction. It takes experience to get this right because not all trees are alike, and carrying arms full of rope around the tree has never been much fun. **Friction devices** have been designed for tree work, and they have some obvious control advantages over taking wraps around the tree. Examples include the various designs of **bollards,** which are posts that strap to the tree, for taking wraps in a **load line**. The large diameter provides a favorable **bend ratio**, which minimizes strength loss in the rigging line. Some bollard-type lowering devices are designed with a ratcheting system that allows the arborist to remove slack from the line or even lift a load. Other friction devices have been designed for smaller loads (Figure 15.26).

Rather than having to tie a knot each time a rope is used, connecting links such as carabiners, **shackles**, or **screw links** can speed the process. Most connectors are not designed for **dynamic loading**, which eliminates their use in most trunk and top

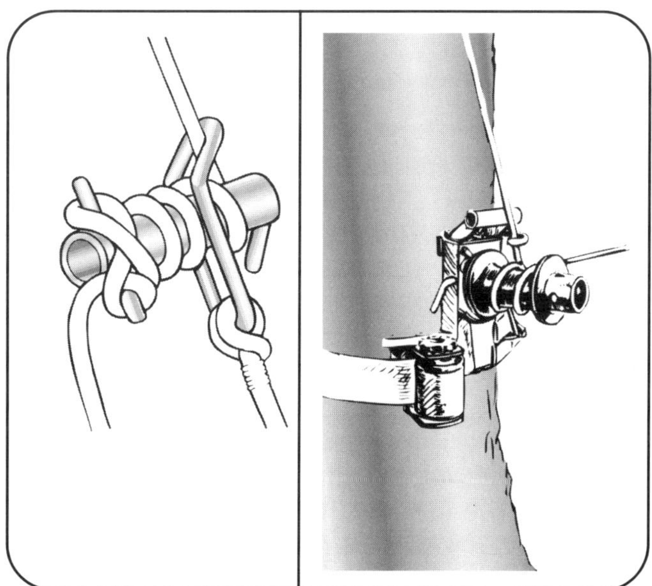

Figure 15.26 Friction devices used for lowering. Port-a-Wrap (left); Hobbs lowering device (right).

removal operations. When used with ropes of ½- or ¾-inch diameter, a ⅜-inch carabiner can produce an unfavorable bend in the rope, which can significantly weaken the system. While aluminum is preferred for climbing because of its light weight, steel has far better strength and fatigue properties. With high loads, many loading cycles, and the potential for dynamic loading, steel connecting links are usually the preferred choice in rigging.

Carabiners must always be loaded along their major axis, and never across the gate. Selecting the correct shape can aid in keeping rope and other devices properly positioned in the carabiner. Common shapes for carabiners are oval, D, modified D, and pear. Some have a fixed eye in addition to the opening formed by the gate, similar to a locking rope snap (Figure 15.27).

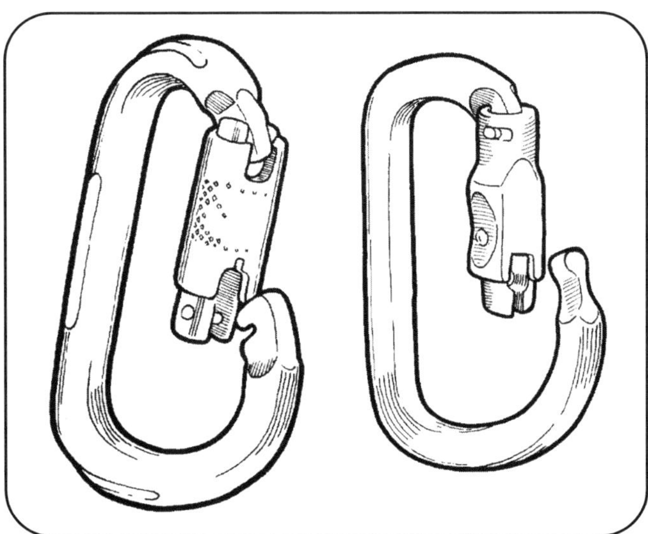

Figure 15.27 Two types of carabiners. Twist-lock with notch gate (left); ball-lock carabiner with key-lock gate (right).

Compared to running lines through tree crotches, the use of **blocks** can decrease dynamic loading, wear on ropes, or damage to the tree. The rotating **sheaves** of pulleys can be mounted with bearings or bushings to reduce friction. Bearings have lower friction but are more susceptible to damage from dirt, and they are not considered as strong in dynamic loading situations. Thus, most **arborist blocks** are made with bushings, while **rescue pulleys** often have bearings. Arborist blocks are heavy-duty pulleys, with a large, rotating sheave for the lowering line, and a smaller, fixed sheave to accept a **rope sling**. These devices are designed specifically for arborists and are descendants of large, industrial snatch blocks. Most significantly, the side plates of an arborist block extend beyond the sheaves to protect the line from abrasion. In contrast, rescue pulleys are designed for static, overhead rigging, where the loads are known and very low friction is required (Figure 15.28).

Through knotting or splicing, rope can be made into any number of tools. There are many variations of slings and other rope tools used in rigging. In addition, **webbing slings** can be purchased in differently sized sewn loops, or they can be knotted from tubular webbing. The strength of a sling depends on the material, the way the loop is formed (sewn or knotted), and the way the loop is used. For example, if the strength in a straight pull is 100 percent, then a choker application (girth hitch) retains about 80 percent of that strength, and a basket hitch, where an object is cradled in a **bight** by both legs of the sling, is twice as strong.

Fundamental Rigging Techniques

There is always more than one way to get the wood and brush to the ground, but the best method is the one that maximizes productivity while still maintaining safety. An investment in equipment, and education into the science behind the equipment, can pay off by making the work easier, reducing wear on tools, and allowing larger sections to be safely removed.

The first choice to be made is whether to run ropes through natural crotches or through a false crotch block. Natural crotches can be fast and effective, but the consistent friction and versatility of placement of false crotches are often great advantages. Just as important as how friction is reduced at the rigging point is how it is added to the rigging

Figure 15.28 Arborist blocks attached to a tree with a rope sling. Using a timber hitch (left); using a cow hitch (right).

system when lowering larger pieces. Certainly, wrapping the rigging line around a tree trunk will work, but the amount of friction added is inconsistent, and the wear on ropes and trees can be excessive. **Lowering devices** offer alternatives to these drawbacks, with the additional ability to help in raising wood.

Choosing an appropriate knot to tie off a section of wood can be important for large pieces, or when dynamic loading will be a concern. In cases where pieces are rigged from above, and swing is controlled, security is still important, but ease of tying and untying can also be considered. One standard knot is the clove hitch. If unsecured, a piece can roll out of the clove hitch, the same way a tautline hitch can roll, so a tail must be left in the working end of the rigging line and at least two **half hitches** tied around the standing part of the line. Another option for tying off pieces is the **running bowline**. The advantage of the running bowline is that the rigging line can be set in a distant crotch, the **working end** retrieved, and the knot secured to that crotch by pulling on the **standing part** of the line.

Techniques of rigging are the different ways to remove wood from a tree. As unsophisticated as it may sound, perhaps the most common technique in pruning is to cut a piece and throw it to a safe **landing zone** (**drop zone**) on the ground. Once ropes are employed, techniques are categorized by the position of the rigging point relative to the work. With the rigging point above the work, most often the piece is tied at the butt end, cut, and lowered to the ground. This technique is appropriate in many situations, and in day-to-day work may be all that is required.

Controlling the swing is important for the safety of the climber and surrounding property. At times, it is not worth the time to piece out a large limb by **butt-tying** several sections, compared to removing one piece with a more complex technique. In these cases, still with the rigging point above, the piece could be tied at the tip and dropped or **tip-tied** and lifted. When the piece must be removed without dropping either the butt or tip, it can be tied so it is **balanced**, then lowered to the ground (Figure 15.29).

With any of these techniques, a **tagline** can (and often should) be used to direct the piece or control its swing. A tagline is a second rope tied to the piece and controlled by a ground worker, but which does not support any of the load. Another technique, termed **redirect rigging**, can be applied when the rigging point is above. Here a single line is run through several anchor points in the tree. This can help distribute forces throughout the canopy, control the piece's swing, or land the brush in a specific location.

There are situations where a rigging line cannot be anchored above the work, such as in removals after all the brush has been removed and the trunk remains. Most common in these cases is **butt-hitching**, where a piece is tied above a cut, and the line is run through a block or crotch below the cut. This can be one of the most demanding techniques for a rope because the possibility of **shock-loading** is so high. Because there are no high tie-in points, it can also be dangerous for the climber (Figure 15.30).

One of the more complex techniques an arborist can apply is the **speedline**. It may not be used very often, but when it is applied it can greatly improve productivity. The simplest form of this technique uses a line stretched from the canopy of a tree to an anchor on the ground. As pieces are cut, they are attached to the speedline and allowed to run to the ground. It is very important to avoid dropping loads onto a speedline. The speedline should be slacked before loading and tightened for running the pieces down. While it may be fast and easy to rig, the technique implemented in this way has some significant drawbacks. Coupling the speedline with another technique, such as butt-tying or butt-hitching, gives a separate line to control the dynamic loading, which can then be used to control the piece's descent on the speedline.

Cutting Techniques

Once the limb is rigged for removal, the climber must decide on the appropriate method of cutting the limb. The **drop cut**—the classic 3-point cut—dates back to the early years of arboriculture and appears in almost every pruning text as the recommended technique for removing large limbs. It consists of an undercut and a top cut farther out on the limb. When using a chain saw, arborists should form the top cut directly above the undercut, to avoid getting the bar stuck in the **kerf** of the cut as the limb breaks free.

A cut that is handy for controlling relatively small sections of wood that may not require roping is the **snap cut**. This cut is made by cutting slightly more than halfway through a section from the side, then cutting from the opposite side, but an inch or more offset from the first cut. The distance apart will need to be larger for larger limbs. The two cuts will bypass, but the fibers should hold. The saw can be

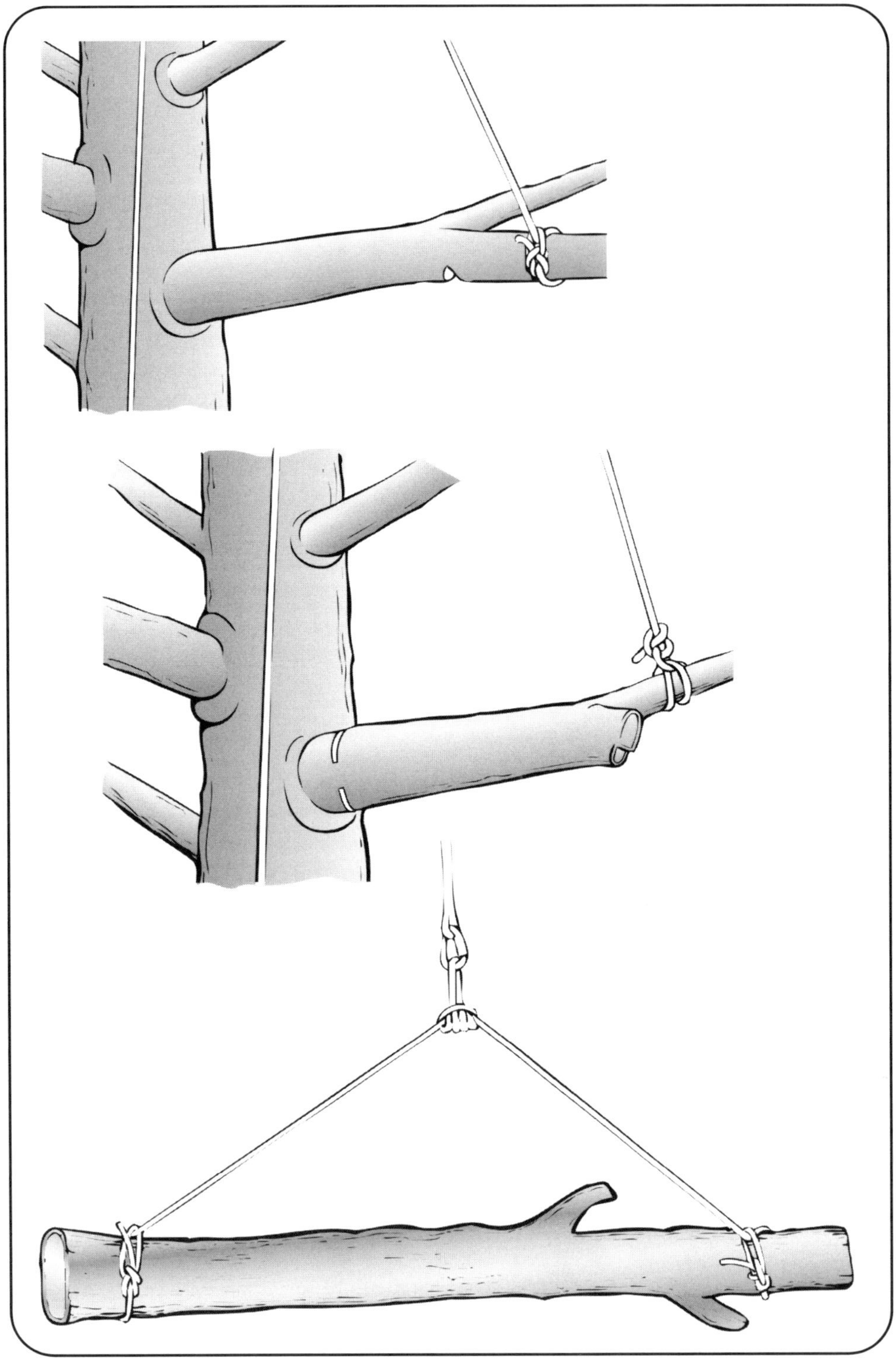

Figure 15.29 Basic rigging techniques. A tagline could be added to any of these for extra control. Butt-tying (top); tip-tying (center); balancing (bottom).

shut off and the remaining piece broken off manually. A variation of this cut can be used to remove the final stub from a large branch that has been removed.

The **hinge cut** is a variation of standard tree-felling techniques. It employs the use of a **notch** and **back cut** to form a **hinge** and "steer" the limb. It can be used to swing a limb around rather than simply dropping the branch to the ground. Unless the limb is supported with a rigging line, there is a limit to how much the climber will be able to swing it before the hinge breaks. If the hinge is formed too far around the side of the limb, the hinge may be ineffective and may break before the limb swings.

The strategy for piecing out a tree depends on the circumstances. The climber must plan the order of removal to avoid being left with a limb that is too difficult or dangerous to remove. A general rule of thumb is to clear a pathway for the limbs, removing brush first. Removing the easiest limbs first can sometimes cause problems when

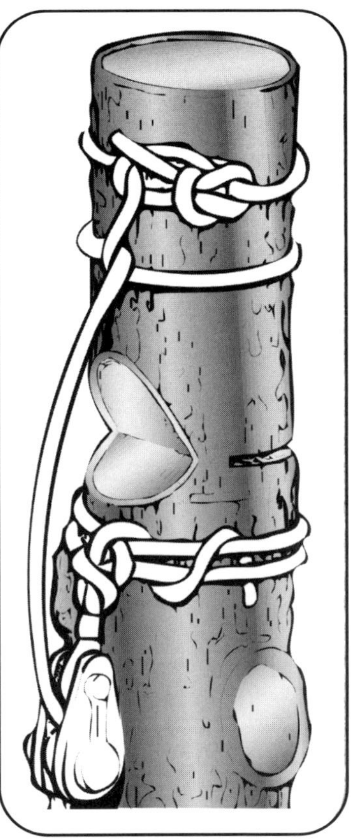

Figure 15.30 Butt-hitching, or blocking out, using an arborist block for the rigging point below the cut.

trying to rig limbs later. As with all aspects of removal, the key to safety is to use equipment that can handle the load. Workers must avoid trying to take out a section that is too big, and must always consider what could happen if some component of the rigging fails. The climber and all ground workers must be clear of danger in the event of something breaking unexpectedly.

Chapter 15 Workbook

1. All equipment used in tree care must conform to and be inspected according to ____________________ ________________.
2. Before climbing a tree, a climber should inspect the tree for hazards such as:
   a.
   b.
   c.
   d.
   e.
3. ___________ __________ are the knots used by climbers to tie in. Three commonly used examples are:
   a.
   b.
   c.
4. A ____________ ___ _____ is often tied in the tail from the climbing hitch as a ___________ _______ to prevent the end from going through the climbing hitch.
5. A ________ is a type of knot used to secure a rope around an object, another rope, or the standing part of the same rope. A __________ joins two rope ends together.
6. Name two methods for installing a climbing line into a tree:
   a.
   b.
7. Two methods of using a climbing line to ascend a tree are ________ _________________ and _________ ______________.
8. ____________ ________________ is a technique in which a climber ties into two places to facilitate climbing or working in a tree.
9. For added stability and safety, a climber should be secured with a ________________ __________ in addition to the climbing line when using a chain saw in a tree.
10. The first step in assessing an emergency situation in a tree is to determine whether there is an ________________ ________________.
11. _____________ is the use of ropes and other equipment to take down trees or remove limbs.
12. ____________ ________ ropes are not recommended for natural crotch rigging, where the friction of the cover with the tree causes an imbalance in the load taken by the core and cover braids.
13. Dividing the __________ ________ of a rope into the tensile strength provides the working load limit.
14. Carabiners must always be loaded along their _________ _______, and never across the gate.
15. _____________ ________ are heavy-duty pulleys with a large, rotating sheave for the lowering line, and a smaller, fixed sheave to accept a rope sling. In contrast, ____________ _________ are designed for static, overhead rigging, where the loads are known and very low friction is required.
16. Natural crotches can be fast and effective for use as a rigging point, but the consistent friction and versatility of placement of a ___________ ___________ is often a great advantage.

17. When the piece must be removed without dropping either the butt or tip, it can be tied so it is _______________, then lowered to the ground.
18. A technique in which a piece is tied above the point where it will be cut, and the line is run through a block or crotch below the cut is called _________ _____________.
19. The ________ ______ is the classic 3-point cut, which dates back to the early years of arboriculture.
20. The _________ ______ is a variation of standard tree-felling techniques that employs the use of a notch and back cut to form a hinge and "steer" the limb.
21. Label the following knots:

______________________ ______________________ ______________________ ______________________

MATCHING

___ ANSI Z133.1	A. may be used to attach a false crotch
___ scabbard	B. used for secured footlocking
___ lowering device	C. safety standards for tree care
___ Prusik loop	D. climbing technique
___ Blake's hitch	E. eliminates need for wraps on the tree in rigging
___ rope sling	F. second climbing line for emergency rescue
___ access line	G. climbing hitch
___ body thrust	H. sheath for a handsaw

CHALLENGE QUESTIONS

1. Describe the inspection process that every climber should follow before climbing a tree. Include inspection of gear and inspection of the tree.

2. What safety precautions should a climber take while ascending a tree? What techniques may be used?

3. Describe the steps involved in rigging large limbs for removal. How can specialized equipment be used to make the job easier and safer?

SAMPLE TEST QUESTIONS

1. "Body thrust" refers to
   a. the safest method of lifting heavy objects
   b. the stage of insect development when the skin is shed to allow for growth
   c. a method of ascending a tree
   d. a physiological problem leading to stress cracks
2. A Prusik loop is used
   a. to tie in with the secured footlock method
   b. to attach limbs to a speedline
   c. as a false crotch to lower limbs
   d. to cable small, multi-stemmed trees
3. A false crotch is
   a. the junction of two codominant stems
   b. a crotch that contains included bark, making it a potential hazard
   c. a block or other device that is hung in a tree to create a rigging point for ropes
   d. a point of weak branch attachment created by sucker development near heading cuts
4. Most arborist climbing and rigging lines are made of
   a. nylon
   b. polyester
   c. polypropylene
   d. spectra
5. The classic 3-point cut used to remove limbs is also called the
   a. drop cut
   b. jump cut
   c. hinge cut
   d. topping cut

Other Sources of Information

(See pages v–vi for complete bibliographic information.)

ANSI Z133.1. *Safety Requirements for Pruning, Trimming, Repairing, Maintaining, and Removing Trees and for Cutting Brush.*
ISA. *ArborMaster Video Series I.*
ISA. *ArborMaster Video Series IV.*
ISA/NAA. *Basic Training for Tree Climbers.*
Jepson, 2000. *Tree Climbers' Companion.*
Lilly, 1998. *Tree Climbers' Guide.*

GLOSSARY

GLOSSARY

3-strand rope—type of rope construction in which three strands are twisted together
7-strand, common-grade cable—type of cable; often used to cable trees
12-strand rope—braided rope construction used for climbing and rigging lines; the 12-strand construction is not easily spliceable
16-strand rope—braided rope construction with a cover and a core
abiotic disorders—plant problems caused by nonliving agents
abscissic acid—plant growth substance that triggers leaf or fruit drop
abscission—leaf or fruit drop
abscission zone—area at the base of the petiole where cellular breakdown leads to leaf drop
absorbing roots—fine, fibrous roots that take up water and minerals; most of them are within the top 12 inches (30 centimeters) of soil
absorption—taking up
access line—a second climbing line hung in a tree as a means of reaching a victim in an emergency
access route—means of entering and leaving a property during a construction operation
acclimation—process by which plants and other living organisms adapt physiologically to a climate or environment different than their own
acuminate—leaf shape having an apex the sides of which are gradually concave and taper to a point
acute—disorder or disease that occurs suddenly or over a short period of time
adaptability—genetic ability of plants and other living organisms to adjust or accommodate to different environments
adpressed—in close, tight proximity
adventitious bud—bud that arises from a place other than a leaf axil
aeration—provision of air to the soil to alleviate compaction and improve its structure
aeration system—the set of holes or trenches created in a tree's root area to improve oxygen availability to the roots
aerial device—truck with booms and bucket used to place a worker in proximity to a tree's crown
aerial rescue—method used to bring an injured worker down from a tree or aerial lift device
aesthetic—artistic or pleasing characteristics
aggregate—close cluster or mix
air excavator—device that blows air at high force; used to remove soil from the root zone of trees
air terminal—the uppermost point of a lightning protection system for trees
allelochemicals—naturally produced substances in plants that serve as part of the plant's defense against pests and that may have effects on the growth and development of other plants
allelopathy—chemical effect or inhibition of the growth or development of plants induced by another plant
alternate—having leaves situated one at each node and alternating positions on the stem; this arrangement means the leaves are not across from each other
amon-eye nut—specialized nut used in cabling trees; has a large eye for attachment of the cable
anatomy—study of the structure and composition (of plants)
angiosperm—plants with seeds borne in an ovary; consisting of two big groups, monocotyledons and dycotyledons
anion—an ion that carries a negative charge
annual rings—see *growth rings*
ANSI A300 standards—industry-developed standards of practice for tree care; acronym for American National Standards Institute
ANSI Z133.1— safety standards for tree care operations
anthocyanins—red, purple, or blue pigments; responsible for those colors in some parts of trees and other plants
antigibberellin—plant growth regulator that inhibits the action of the plant hormone gibberellin
antitranspirant—substance sprayed on plants to reduce water loss through the foliage
apical—having to do with the tip
apical bud—terminal bud on a stem
apical dominance—condition in which the terminal bud inhibits the growth and development of lateral buds on the same stem
apical meristems—the growing points at the tips of shoots
appropriate response process (ARP)—method of systematically assessing plant health and client needs to determine which course of action, if any, is recommended
approved—acceptable to federal, state, provincial, or local enforcement authorities

arboriculture—the study of trees and other plants
arborist block—heavy-duty pulley with two attachment points and extended cheek plates; used in rigging operations
artificial respiration—forcing air into the lungs of a person who has stopped breathing
auxin—plant hormone or substance that promotes or regulates the growth and development of plants; it is produced at sites where cells are dividing, primarily in the shoot tips
available water—the water remaining in the soil after gravitational water has drained and before the wilting point has been reached
axial transport—movement of water, minerals, or photosynthate longitudinally within a tree
axillary bud—bud in the axil of a leaf; lateral bud
back cut—cut made on opposite side of a log toward the notch cut or face cut
backfill—soil (and amendments) put back into the hole when planting a tree
balance—in rigging, a technique used to lower a limb without allowing either end to drop
balled and burlapped (B & B)—having the root system and soil wrapped in burlap for moving and planting a tree or other plant
barber chair—dangerous condition created when a tree or branch splits vertically up from the back cut
bare root—tree or other plant taken from the nursery with exposed root system, without soil
bark—protective covering over branches and stem that arises from the cork cambium
bark tracing—cutting away torn or injured bark to leave a smooth edge
barrier—fences or other means used to establish a protection zone around trees on construction sites
belay—securing a climber's rope using wraps around a cleat, carabiner, or other device
bend—type of knot used to join two rope ends together
bend ratio—ratio of the diameter of a branch, sheave, or other device to the rope that is wrapped on it
bight—a curve or arc in the active part of a rope between the working end and the standing part
biodegradable—capable of decaying and being absorbed by the environment
biological control—method of controlling plant pests through the use of natural predators, parasites, or pathogens
biotic—pertaining to a living organism
biotic disorders—disorder caused by a living agent
bipinnate—double pinnate; see *pinnate*
blade—the expanded body of a leaf
Blake's hitch—a climber's friction knot sometimes used in place of the tautline hitch or Prusik knot
blight—any disease, regardless of the causal agent, that kills young, plant-growing tissues
block—a pulley used in rigging
body thrust—method of ascending a tree using a rope
bollard—a post on which wraps can be taken with a rope
boom—long, movable arm of a bucket truck
botanicals—pesticides that are made from plants
bowline—looped knot used to attach items to a rope
bowline on a bight—knot that can be used as a makeshift saddle in an emergency situation
box cable system—tree cabling system that forms closed polygons; used to join together more than three branches
bracing—installation of metal rods through weak portions of a tree for added support
bracing rod—metal rod used to support weak sections or crotches of a tree
branch bark ridge—top area of a tree's crotch where the growth and development of the two adjoining limbs push the bark into a ridge
branch collar—area where a branch joins another branch or trunk created by the overlapping xylem tissues
branch protection zone—tissues inside the trunk or parent branch at the base of a subordinate branch that protect against the spread of decay
branch union—point where a branch originates from the trunk or another branch; crotch
broadcast fertilization—application of fertilizer over the soil surface
bud—small lateral or terminal protuberance on the stem of a plant that may develop into a flower or shoot; undeveloped flower or shoot
buffering capacity—ability of a soil to maintain its pH
bulk density—mass of soil per unit volume; often used as a measure of compaction
butt-hitching—method of lowering pieces when the rigging point is below the work
butt-tying—tying off a limb at the butt end for rigging
buttress roots—roots at the base of the trunk; trunk flare
CPR—cardiopulmonary resuscitation
CSA—Canadian Standards Association
cable aid—device used to tighten lags and to aid in cable installation
cable clamp—a double-bolted, U-shaped clamp used to secure tree cables
cable grip—device used to attach extra-high-strength cable to lag hooks or eye bolts

cabling—installation of hardware in a tree to help support weak branches or crotches

cambium—layer(s) of meristematic cells that give rise to the phloem and xylem and allow for diameter increase in a tree

canker—localized diseased area, often shrunken and discolored, on stems and branches

carabiner—oblong metal ring used in climbing and rigging that is opened and closed by means of a spring-loaded gate

carbohydrate—compound, combining carbon and water, produced by plants during photosynthesis

cardiopulmonary resuscitation—procedure used to force air into the lungs and to force blood circulation in a person who has suffered cardiac arrest

carotenoid—a yellow, orange, or red pigment responsible for those colors in some parts of trees and other plants

cation—a positively charged ion

cation exchange capacity (CEC)—ability of a soil to adsorb and hold cations

cavity—an open wound or hollow within a tree, usually associated with decay

cell—smallest unit of an organism that is capable of self-reproduction

cellulose—complex carbohydrate found in the cellular walls of the majority of plants, algae, and certain fungi

central leader—the main stem of a tree, particularly an excurrent specimen

chaps—a form of leg protection; worn when operating chain saws

chlorophyll—green pigment of plants, found in chloroplasts; it captures the energy of the sun and is essential in photosynthesis

chloroplast—specialized organelle found in some cells; the site of photosynthesis

chlorosis—whitish or yellowish discoloration caused by lack of chlorophyll; often used in referring to a plant's foliage

chronic—disorder or disease occurring over a long period of time

class—taxonomic group below the division level but above the order level

climbing hitch—knot used as the primary friction knot (tie-in knot) in climbing

climbing line—a rope that meets specifications for use in tree climbing

climbing saddle—a harness designed for climbing trees

climbing spurs—sharp devices that can be strapped to a climber's lower legs to assist in climbing poles or trees being removed

clone—asexually produced organisms that are genetically identical

clove hitch—knot used to secure an object to a rope

CODIT—Compartmentalization Of Decay In Trees

codominant branches/codominant stems—forked branches of nearly the same size in diameter and lacking a normal branch union

come-along—portable cable winch used to draw two things closer together

command and response system—a system of vocal communication convention used in tree care operations

compaction—compression of soil resulting in the loss of macropores

companion cell—parenchyma cells associated with sieve tube members

compartmentalization—natural process of defense in trees by which they wall off decay in the wood; see *CODIT*

complete fertilizer—fertilizer that contains nitrogen, phosphorus, and potassium

complex—a combination of factors that contribute to the stress or decline of a tree

compound leaf—a leaf with two or more leaflets

conduction—carrying water or nutrients

conifer—a cone-bearing tree or other plant that has its seeds in a structure called a cone

conk—the fruiting body of a fungus, often associated with decay

contact insecticides—materials that cause injury or death to an insect after coming in contact with the pest

container grown—tree or other plant that has been grown in a container

containerized—plant available from the nursery with its root mass in a container

controlled-release fertilizer—slow-release or slowly soluble form of fertilizer

conventional notch—45-degree notch with a horizontal bottom cut; used in felling trees

cordate—heart-shaped

cork cambium—meristematic tissue from which cork and bark develop to the outside

cracks—defects in trees that, if severe, may pose a risk of tree or branch failure

crenate—leaf margin with rounded teeth

cross section—section cut perpendicular to the axis of longitudinal growth

crown—the aboveground portions of a tree

crown cleaning—removal of watersprouts and dead, dying, diseased, crossing, and hazardous branches from a tree

crown reduction—method of reducing the height or spread of a tree by performing appropriate pruning cuts

crown restoration—method of restoring the natural growth habit of a tree that has been topped or damaged in any other way

crown rot—disease or decay at the base of a tree or root flare

cultivar—a cultivated variety of a plant

cultural control—method of controlling plant pests by providing a growing environment favorable to the host plant and/or unfavorable to the pest

cuticle—waxy layer outside the epidermis of a leaf

cycles to failure—number of times a rope or other piece of equipment can be used before mechanical failure

cytokinin—plant hormone involved in cell division

D-rings—D-shaped metal rings on a climber's saddle used to attach ropes and snaps

dead-end grips—cable termination devices that must be used with extra-high-strength cable

dead-end hardware—cabling, bracing, or guying hardware that is terminated by screwing into the tree

deadwooding—removal of dead and dying limbs from a tree

decay—decomposition of woody tissues by fungi or bacteria

deciduous—tree or other plant that loses its leaves sometime during the year and stays leafless generally during the cold season

decurrent—rounded or spreading growth habit of the crown of a tree

deficiency—lack or insufficient quantity of a required element

defoliation—loss of leaves from a tree or other plant by biological or mechanical means

dentate—having marginal teeth that are perpendicular to the leaf margin

desiccation—total drying out

design criteria—aspects of the site and required functions to be served by the plant that must be considered in plant selection

design factor—factor by which the tensile strength of a rope or piece of hardware is reduced to arrive at the working-load limit for a given application

dicot—see *dicotyledon*

dicotyledon—plant with two cotyledons in its embryo

dieback—condition in which the ends of the branches are dying

differentiation—process in the development of cells in which they become specialized for various functions

diffuse porous—pattern of wood development in which the vessels are distributed evenly throughout the annual ring

dioecious—plant with unisexual flowers with each sex confined to separate plants

direct cable system—simple tree cabling system to join two branches

direct contact—when any part of the body touches an electrical conductor

division—taxonomic division below kingdom level but above class level

dormant—state of reduced physiological activity in the organs of a plant

double braid—rope construction that consists of a braided rope within a braided rope

double crotch—climbing technique consisting of tying into two places in a tree

double serrate—toothed margin of a leaf with smaller teeth within

downy mildew—white fungal growth emerging from water-soaked tissue, usually on the underside of the leaf

drill-hole method—applying fertilizer by drilling holes in the soil occupied by the roots or surrounding them

drip irrigation—method of watering in which water evaporation and runoff are minimized

drip line—perimeter of the area under a tree delineated by the crown

drop-crotch pruning—method of reducing the height of a tree; see *reduction*

drop cut—branch-removal technique consisting of an undercut and a top cut farther out on the branch

drop zone—area where cut branches or wood sections will be dropped from a tree

drum lace—method of tying a balled-and-burlapped tree root ball for moving

dynamic loading—forces created by a moving load; load that changes with time

electrical conductor—body or medium that allows the passage of electricity; while working on trees, generally this will be any overhead or underground electrical device, including communication cables and power lines that have electricity or the potential to have it

emergency response—predetermined set of processes by which emergency situations are assessed and handled
entire—leaf margin without teeth
epicormic—arising from latent or adventitious buds
epidermis—outer tissue of leaves, stems, roots, flowers, and seeds
epinasty—distortion of growth
espalier—specialized technique of pruning and training plants to grow within a plane
essential elements—the 17 minerals essential to the growth and development of trees
ethylene gas—naturally occurring plant growth substance that triggers fruit ripening
evapotranspiration (ET)—moisture lost by evaporation of the soil's water and transpiration of the plant
evergreen—tree or plant that keeps its needles or leaves year round; this means for more than one growing season
excurrent—tree growth habit with pyramidal crown and a central leader
exfoliating—peeling off in shreds or layers
extra-high-strength cable—type of cable used in supporting trees; stronger but less flexible than standard wrapped cable
exudation—oozing out
eye bolt—a drop-forged, closed-eye bolt installed in trees to attach cable
eye splice—technique used to attach common-grade cable to eye bolts or lags
eyesplice—termination in a rope forming an eye and made by splicing the rope back upon itself
face cut—a notch cut used in felling trees or limbs
fall protection—equipment and techniques designed to ensure a climber will not fall from a tree
false crotch—device installed in a tree to set ropes during climbing or rigging when there is not a suitable natural crotch available
family—the taxonomic division under the order level and above the genus level
fermentation—incomplete path of respiration in the absence of sufficient oxygen
fertilizer—substance added to a plant or the surrounding soil to supplement the supply of essential elements
fertilizer analysis—the percentage of nitrogen, phosphorus, and potassium in a fertilizer
fertilizer burn—injury to plants resulting from excess fertilizer salts in the surrounding soil
fiber—elongated, tapering, thick-walled cell that provides strength
field capacity—soil moisture content following the drainage of gravitational water
figure-8 descender—metal device used in rigging
figure-8 knot—safety knot or stopper tied in the climbing line
first aid—emergency care or treatment of the injuries or illnesses of a person to stabilize his or her condition before medical help is available
foliage—the leaves of a plant
foliar analysis—laboratory analysis of the mineral content of foliage
foliar application—application of a fertilizer or other substance by direct spray on the foliage
footlocking—method of climbing a rope by wrapping the rope around one's feet
friction device—device used to take wraps in a load line; provides friction for controlled lowering
friction hitch—any of several friction knots used in climbing trees or rigging
fronds—large, divided leaves, as in palms
fruiting bodies—the reproductive structures of fungi, the presence of which may indicate decay in a tree
fungicides—chemical compounds that are toxic to fungi
gall—swelling of plant tissues; frequently caused by insects, nematodes, fungi, or bacteria
genus—a group of species having similar fundamental traits; botanical classification under the family level and above the species level
geotropism—plant growth produced as a response to the force of gravity; it can be positive as in the roots, or negative as in the trunk
gibberellin—a plant growth substance involved in cell elongation
girdling—inhibition of the flow of water and nutrients in a tree by choking vascular elements
girdling root—root that grows around a portion of the trunk of a tree, causing inhibition of the flow of water and nutrients by choking vascular elements
gravitational water—water that drains from the soil's larger macropores under the force of gravity
ground rod—10-foot (3-meter) metal rod used in grounding a lightning protection system
grounded—electrically connected to the earth
growth rate—speed at which something grows
growth rings—rings of annual xylem visible in a cross section of the trunk of some trees
guard cells—pair of cells that regulate the opening and closing of a stomate due to a change in water content
gummosis—exudation of sap or gum, often in response to disease or insect damage

guying—securing a tree, if needed, with ropes or cables fastened to anchors in the ground or another tree
gymnosperm—plant with seeds exposed
half hitch—simple wrap of a rope used to secure a line temporarily
hardened off—acclimated to the cold or to a new environment
hardiness—ability of a plant to survive low temperatures
hazard assessment—process by which the risk potential of a tree is determined
hazard potential—degree of risk posed by a tree
heading back—topping; cutting limbs back to buds, stubs, or lateral branches not large enough to assume apical dominance
heartwood—inner, nonfunctional xylem tissues that provide structural resistance to the trunk
hinge—a strip of wood fibers created between the notch and the back cut that help control direction in tree felling
hinge cut—sequence of cuts used to control the direction of a limb being removed
hitch—a knot made when a rope is secured around an object or its own standing part
hollow braid—rope construction characterized by a braided rope with no core
honeydew—substance secreted by certain insects when feeding upon plants
horizon—layer of soil within the soil profile
horticultural oils—highly refined petroleum oils used to smother insects and disrupt their membranes
Humboldt notch—a felling notch that is horizontal on the top and angled on the bottom
IPM—see *Integrated Pest Management*
identification key—aid used to help identify plants
implant—device, capsule, or pellet that can be inserted into a tree to treat disorders
included bark—bark that becomes embedded in a crotch between branch and trunk or between codominant stems and causes a weak structure
increment borer—device used to take core samples from trees for the purpose of determining age or detecting problems
indirect contact—touching any conductive object that is in contact with an electrical conductor
infectious—capable of being spread from plant to plant
infiltration—downward entry of water into the soil
infiltration rate—speed at which water soaks into the soil
inorganic fertilizer—mineral fertilizer, not coming from plant or animal
insect growth regulators—substances, naturally occurring in insects, that affect growth and development
insecticidal soaps—mild salts of fatty acids that disrupt insect life processes
insecticides—substances that are toxic to insects
Integrated Pest Management (IPM)—method of controlling plant pests combining biological, cultural, and chemical controls
internodal—between the nodes on a stem
internode—the region of the stem between two successive nodes
interveinal tissue—leaf tissue between the veins or vascular bundles
introduced species—plant species that are not native to a region
ion—one atom or a group of atoms with a positive or negative charge
job briefing—brief meeting of a tree crew at the start of every job to communicate the work plan, responsibilities and requirements, and any potential hazards
kerf—slit or cut in a log made by a saw
kernmantle—rope manufactured to have a core and woven sheath
key—plant identification tool used to determine a plant species
kickback—sudden backward or upward thrust of a chain saw
kickback quadrant—upper quadrant of the tip of a chain saw bar
kingdom—the primary division in taxonomy, separating plants from animals
lag eye—lag-threaded cable anchor with a closed eye
lag hook/J-hook—J-shaped bolt used to attach cables to trees
lag-threaded rod—steel bracing rod used to screw into a predrilled hole to provide added support to a tree
landing zone—predetermined area where parts will be brought down in a rigging operation
landscape function—the environmental, aesthetic, or architectural functions that a plant can have
lanyard—a short rope equipped with snaps or carabiners; work-positioning lanyards are used for temporarily securing a climber in one place
larva—immature life stage of an insect
lateral—secondary or subordinate branch
lateral bud—vegetative bud on the side of a stem
lateral root—side-branching root that grows horizontally

leach/leaching—tendency for elements to wash down through the soil
leader—the primary terminal shoot or trunk of a tree
leaf apex—tip of the leaf blade
leaf base—bottom part of the leaf blade
leaf blotch—irregularly shaped areas of disease on plant foliage
leaf margin—outer edge of the leaf blade
leaf scar—scar left on the twig after a leaf falls
leaf spot—patches of disease or other damage on plant foliage
leaflet—separate part of a compound leaf blade
leg protection—chaps or other protective clothing worn over the legs when operating a chain saw
lenticel—opening in the bark that permits the exchange of gases
liability—something for which one is responsible; legal responsibility
lignin—substance that impregnates certain cell walls
lion tailing—poor pruning practice in which limbs are thinned from the inside of the crown to a clump of terminal foliage
liquid injection—method of injecting liquid forms of fertilizer into the surrounding soil of a tree
load line—rope used to lower a tree branch or segment that has been cut
lobe—projecting segment of a leaf blade
lowering device—instrument attached to the base of a tree in rigging; used to take wraps with the load lines
machine-threaded rod—steel rod used in cabling and bracing; must be terminated with washers and bolts
macronutrient—any of the essential elements required by plants in relatively large quantities
macropore—larger spaces between soil particles that are usually air-filled
main conductor—primary conductor cable of a lightning protection system; standard down conductor
mature height—the maximum height that a plant can reach if the conditions of the planting site are favorable
meristem—undifferentiated tissue in which active cell division takes place
microbial extracts—substances derived from microorganisms; may be used to control certain pests
microinjection—method used to introduce chemicals directly into the xylem of trees
micronutrient—any of the essential elements required by plants in relatively small quantities
micropore—space between soil particles that is relatively small and likely to be water filled
micropulley—small pulley used by tree climbers
minimum irrigation—the practice of minimal irrigation through the use of drought-tolerant plants and watering only when necessary due to reduced rainfall
mismatch cut—cut type used in branch removal in which offset, overlapping cuts allow the section to be manually broken off; snap cut
mitigation—process of reducing damages or risk
monitoring—keeping a close watch; performing regular checks or inspections
monocot—see *monocotyledon*
monocotyledon—a plant whose embryo has one seed leaf (cotyledon)
monoecious—a plant with both sexes on the same plant
morphology—study of the form and structure of living organisms; in this case, of plants
mortality spiral—sequence of events causing the decline, and eventual death, of a tree
mycorrhizae—a symbiotic association between a fungus and the roots of a plant
native species—indigenous to a region
naturalized species—a non-native species that has become established in a region
necrosis—localized death of tissue in a living organism
needle—slender conifer leaf
negligence—failure to exercise due care
nematode—microscopic roundworm; some feed on plant tissues and may cause disease
node—slightly enlarged portion of a stem where leaves and buds arise
nomenclature—scientific naming system for living organisms; scientific names are written in Latin, the genus first (starting with capital letter), followed by the species (always with lowercase letter)
notch—a wedge cut into a log or tree for felling
nutrient cycling—movement of mineral nutrients within an ecosystem as organic matter decomposes and is recycled in plants
oblique—lop-sided, one side larger than the other
obtuse—rounded, approaching semi-circular
Occupational Safety and Health Act (OSHA)—in the United States, the legislative act dealing with health and safety in the work place; administered by the Occupational Safety and Health Administration; Occupational Health and Safety Administration (OHSA) in Canada
oedema (edema)—watery swelling in plant tissue
oils—highly refined petroleum oils used to smother insects and disrupt their membranes

open-face notch—wedge-shaped cut (commonly about 70 degrees) used in felling trees or removing tree sections

opposite—opposite leaf arrangement: leaves situated two at each node, across from each other on the stem

order—taxonomic division below class level but above family level

organic fertilizer—fertilizer derived from plants or animals

organic layer—layer of organic matter at the soil's surface

osmosis—diffusion of water through a semi-permeable membrane from a region of higher water potential to a region of lower water potential

outriggers—projecting structures on boom trucks and other large vehicles; used for stabilization

pH—a measure of acidity or alkalinity of a medium

palmate—radiating in a fanlike manner; type of compound leaf

parasite—organism living in or on another organism from which it derives nourishment

parenchyma cells—thin-walled, living cells essential in photosynthesis and storage

parent material—soil bedrock material from which the soil's profile develops

pathogen—causal agent of disease

perched water table—accumulation of water in an upper soil layer

percolation—movement of water through the soil

permanent branches—branches that will be left in place, often forming the initial scaffold framework of a tree

permanent wilting point—point at which a plant cannot pull any more water from the soil

personal protective equipment (PPE)—personal safety gear such as hard hat, safety glasses, and hearing protection

pest resistance—in plants, the tendency to withstand, or not to get, certain pest problems

pesticides—chemicals used to kill unwanted organisms such as weeds, insects, or fungi

pest resurgence—increase in the population of a pest following a reduction in the population of natural predators or parasites of that pest

petiole—the stalk or support axis of a leaf

petiolule—the stalk of a leaflet

phenols—naturally produced organic alcohols with acidic properties; one of several chemical defense compounds in trees

pheromone—chemical substance produced by insects that serves as a stimulus to other insects of the same species

phloem—plant vascular tissue that conducts photosynthates; situated to the inside of the bark

photoperiod—length of daylight required for certain developmental processes and growth of a plant

photosynthate—general term for the products of photosynthesis

photosynthesis—the process in green plants (and in some bacteria) by which light energy is used to form organic compounds from water and carbon dioxide

phototropism—influence of light on the direction of plant growth

phylum—primary taxonomic division within a kingdom; the plural is phyla

physiological disorder—in plants, a disorder not caused by an insect, pathogen, or injury

physiology—the study of the life function (of a plant)

phytotoxic—a term to describe a compound that is poisonous to plants

pigment—substance that appears colored due to the absorption of certain light wavelengths

pinnate—compound leaf with leaflets along each side of a common axis

plant growth regulator—a compound, effective in small quantities, that affects the growth and development of plants

plant growth substance—a naturally produced compound, effective in small quantities, that affects the growth and development of plants; see *plant hormone*

Plant Health Care (PHC)—a holistic and comprehensive program to manage the health, structure, and appearance of plants in the landscape

plant hormone—substance produced by a plant that affects physiological processes such as growth; see *plant growth substance*

planting specifications—detailed plans and statements of particular procedures and standards for planting

pole pruner—long-handled tool used to make small pruning cuts that cannot be reached with hand tools

pole saw—long-handled tool with a pruning saw on the end

pollarding—a specialty pruning technique used on large-maturing trees that results in the development of callus at the cut ends of the branches

positive-locking—unable to be opened unintentionally; locks automatically and requires two or more motions before opening

powdery mildew—white or grayish fungal growth on the surface of stems or foliage

preformed tree grip—device used to attach extra-high-strength cable to lag hooks or eye bolts
prescription fertilization—philosophy of basing fertilization recommendations on plant needs
pruning—cutting away unwanted parts of a plant
Prusik hitch—type of multi-wrapped friction hitch used in climbing and rigging; used to attach the Prusik loop to the climbing line when footlocking
Prusik loop—loop of rope, smaller in diameter than the climbing line, used for the secured footlock method of ascending a rope
radial aeration—means of aerating the soil in the root zone of a tree by removing and replacing soil in a spokelike pattern
radial transport—movement of substances in a tree perpendicular to the longitudinal axis of the tree
radial trenching—method of improving aeration in the root zone of a tree; radial aeration
raising—removing lower limbs from a tree to provide clearance
ray—tissues that extend radially across the xylem and phloem of a tree
reaction wood—wood formed in leaning or crooked stems, or on lower or upper sides of branches
reaction zone—a natural boundary formed by a tree to separate wood infected by disease organisms from healthy wood; important in the process of compartmentalization
reactive forces—the forces generated in operating a chain saw
redirect rigging—changing the path of a rigging line to modify the forces or the direction of limb removal
reduction—pruning to decrease height and/or spread of a branch or crown
rescue kit—climbing gear and emergency equipment that should be set out on every job site so that it is available in an emergency situation
rescue pulley—light-duty pulley used in rigging operations
resistance—in plants, the tendency to withstand, or not to get, certain diseases
resistant varieties—plants that are tolerant of, or not susceptible to, certain disease or pest problems
resource allocation—distribution and use of photosynthate for various plant functions and processes
respiration—process by which carbohydrates are converted into energy by using oxygen
restoration—pruning to improve the structure, form, and appearance of trees that have been severely headed, vandalized, or damaged
rhizosphere—immediate environment of roots where biological activity is high
rigging—method of using ropes and hardware to remove large limbs or take down trees
rigging line—rope used in rigging operations; usually the load-bearing line
rigging point—the place in the tree (natural or false crotch) that the load line passes through to control limb removal in rigging operations
ring porous—pattern of wood development in which the large-diameter vessels are concentrated in the earlywood
risk assessment—process of determining the level of risk posed by a tree or group of trees on a property
risk management—process of assessing and controlling risk in tree management
root ball—containment of roots and soil of a tree or other plant
root hair—modified epidermal cells of a root that aid in the absorption of water and minerals
root pruning—in transplanting, the process of pre-digging a root ball to increase the density of root development within the final ball
rope sling—a section of rope, usually with at least one eyesplice, used to secure equipment or tree sections in rigging operations
running bowline—knot often used to tie off limbs for removal
rust—disease caused by a certain group of fungi and characterized by reddish brown spots
sanitation—practice of removing dead or diseased plant parts to reduce the spread of disease
sapwood—outer wood that actively transports water and minerals
scabbard—sheath for a pruning saw
scaffold branches—the permanent or structural branches of a tree
scale—one of a group of insects that attach themselves to plant parts and suck the sap
scorch—browning and shriveling of foliage, especially at the leaf margin
screw link—connecting device with a threaded closure mechanism; used in rigging operations
secondary nutrients—nutrients required in moderate amounts by plants
secondary pest outbreak—increase in a secondary pest population following a reduction in the population of natural predators or parasites
secured footlock—method of ascending a rope in which the climber is secured against falling
serrate—sawtooth margin of a leaf with the teeth pointed forward

shackle—a U-shaped fitting with a pin run through it; clevis
shakes—separation of the growth rings in wood
shall—the word that designates a mandatory requirement in the ANSI standards
sheave—the inner fitting within a block over which the rope runs
ship auger—type of drill bit used to drill holes in trees for cable installation
shock-loading—the dynamic load placed on a rope or rigging apparatus when a moving log is stopped
should—the word that designates an advisory recommendation in the ANSI standards
sieve cells—long, slender phloem cells in gymnosperms
sieve tube elements—specialized phloem cells involved in photosynthate transport
sign—the physical evidence of a causal agent
simple leaf—a single, one-part leaf; not composed of leaflets
sink—a plant part that uses more energy than it produces
sinker roots—downward-growing roots that take up water and minerals; most are in the top 12 inches (30 centimeters) of soil
sinus—space between two lobes of a leaf
site analysis—determination of the conditions, environment, and needs of a planting site
site considerations—the factors that must be taken into account when assessing a planting site to select plant species
skeletonized—leaves that have had the tissue removed from between the veins by insects
sling—device used in rigging to secure equipment or pieces being rigged
slowly soluble fertilizer—fertilizer formulation that is slowly hydrolyzed in the soil
slow-release fertilizer—fertilizer that is at least 50 percent water-insoluble nitrogen (WIN)
snap—connecting device used by tree climbers primarily for connecting the climbing line to the saddle
snap cut—cut type used in branch removal in which offset, overlapping cuts allow the section to be manually broken off; mismatch cut
soil amendment—material added to soil to improve its physical or chemical properties
soil analysis—analysis of soil to determine pH, mineral composition, structure, and other characteristics
soil auger—device for removing cores of soil for inspecting or testing
soil compaction—compression of the soil resulting in a reduction of the total pore space, especially the macropores
soil profile—vertical section through a soil, through all of the horizons
soil structure—the arrangement of soil particles
soil texture—the relative fineness or coarseness of a soil due to particle size
source—plant part that produces carbohydrates; mature leaves are sources
species—a group of organisms composed of individuals of the same genus that can reproduce among themselves and have similar offspring
specific epithet—the classification name that follows the genus name in scientific nomenclature
specifications—detailed plans and statements of particular procedures and standards
speed lining—a method of lowering tree segments past obstacles below
speedline—rigging line strung in such a way as to slide tree segments to the ground
splits—open cracks or fissures in tree trunks or branches
split-tail—tree climbing system in which the climbing hitch is formed with a separate, short length of rope
square knot—a knot used to tie together two ropes of equal diameter
staking—supporting a newly planted tree with stakes
standard down conductor—length of copper cable used in lightning protection systems on trees
standing part—the inactive part of a rope, as opposed to the working end
stippling—speckled or dotted areas on foliage
stomata—small pores between two guard cells on leaves and other green plant parts through which gases are exchanged
stopper knot—knot tied in the end of line to keep the tail from passing through the climbing hitch
stress—factor that negatively affects the health of a tree
structural defects—flaws, decay, or other faults in the trunk, branches, or root collar or a tree, which may lead to failure
structural pruning—pruning to establish a strong branch scaffold system
stunting—growth reduction of organisms, in this case, plants
subordinate—pruning to reduce the size and growth of a branch in relation to other branches or leaders
subsurface application—placement of fertilizer below the soil surface

sucker—shoot arising from the roots
surface application—placement of fertilizer or other material on the soil surface
symbiosis—a mutually beneficial association of two different types of living organisms
symbiotic—a mutually beneficial association
symbiotic relationship—association between two organisms that is mutually beneficial
symptom—a plant's reaction to a disorder
systemic—substance that moves throughout and is absorbed by the entire organism, in this case, by the roots, leaves, or both
tagline—rope used to control the swing and direction of drop of a limb being removed
tannins—organic substances produced by trees; believed to be involved in the tree's chemical defense processes
tap root—central, vertical root that grows right below the trunk and is often choked off by the development of other roots
taper—the change in diameter over the length of trunks and branches
target—person, object, or structure that could be injured or damaged in the event of tree or branch failure
tautline hitch—type of climbing hitch used by climbers to tie in
taxonomy—science that studies the description, denomination, and classification of living organisms, based on their similarities and differences
temporary branches—branches left in place when training young trees; such branches will be removed later
tensile strength—the breaking strength of a rope under load
tensiometer—instrument used to measure soil moisture
terminal bud—the bud on the end of a twig or shoot
terracing—method used to lower the soil grade in stages
thimble—device used in cabling to form the loop in the cable
thinning—selective removal of unwanted branches and limbs to provide light or air penetration through the tree or to lighten the weight of the remaining branches
threaded rod metal rod used for support bracing of trees
thresholds—pest population levels requiring action
through-hardware—anchors or braces that pass completely through a trunk or branch and are secured with washers and nuts
throwing ball—device used to set a rope in a tree
throwing knot—a series of loops and wraps tied in a rope to form a weight for throwing
throwline—device consisting of a small weight attached to a thin, lightweight cord; used to set climbing ropes in trees
tie in—to secure a climber's rope in a tree with a tautline hitch
timber hitch knot consisting of a series of wraps on a rope; used to secure the rope to a limb or tree
tip tying—tying a rope on the tip (brush) end of a limb to be removed
topping—cutting back a tree to buds, stubs, or laterals not large enough to assume apical dominance
torts—wrongful acts, other than breach of contract, for which civil action may be taken
tracheid—elongated, tapering xylem cell, adapted for support
translocated—movement of sugars in the phloem
transpiration—water vapor loss through the stomata of leaves
transplant shock—stress following transplant in which growth is reduced and the tree may wilt or drop foliage
transplanting—moving a plant to a new location
tree island—soil or landscape surrounding a tree, such as within a paved area
tree spade—mechanical device used to dig and move trees
tree well—wall constructed around a tree when the soil grade is raised to maintain the original soil level and provide oxygen to the root zone
tree wrap—material used to wrap the trunks of newly planted trees
trenching—digging to install utilities; of concern due to root damage
triangular cable system—tree cabling that forms a triangular shape
tropism—growth movement or variation of a plant as a response to an external stimulus such as light or gravity
tunneling—alternate means to trenching for installation of underground utilities
turgid—fully hydrated to a normal state of distension
undercut—a cut on the underside of a limb to be removed to prevent unwanted tearing as the limb falls
utility pruning—pruning around or near utility facilities with the object of maintaining safe and reliable utility service
variety—subdivision of a species having a distinct difference, and breeding true to that difference

vascular discoloration—darkening of the vascular tissues of woody plants in response to disease
vascular tissue—tissue that conducts water or nutrients
vector—organism that transmits a pathogen
venation—arrangement of veins
vertical mulching—filling vertical drilled holes in the soil with materials such as gravel, perlite, peat, or sand
vessels—stacked, tubelike, water-conducting cells in the xylem
vigor—overall health; capacity to grow and resist stress
vista pruning—selective pruning to allow a view from a predetermined point
vitality—overall health; a plant's ability to deal effectively with stress
water shoot—a secondary, upright shoot arising from the trunk, branches, or roots of a plant
water-holding capacity—ability of a soil to hold moisture
water-insoluble nitrogen (WIN)—nitrogen fertilizer in a form that is not soluble in water
watersprout—an upright, adventitious shoot arising from the trunk or branches of a plant; although incorrect, it is also called sucker
webbing sling—length of sewn webbing, often formed into a loop, used as an attachment in rigging
whorled—leaves arranged in a circle around a point on the stem
wilt—loss of turgidity and subsequent drooping of leaves
wire basket—type of metal basket used to support the root ball of balled-and-burlapped plants
witch's broom—plant disorder in which a large number of accessory shoots develop
work plan—predetermined, orderly means for job completion
working end—the part of a rope terminated for use
working-load limit (WLL)—tensile strength divided by design factor; load limit for a rope or piece of equipment
work-positioning lanyard—rope or strap designed to aid in climbing and tree work; secondary means of attachment
wound dressing—compound applied to tree wounds or cuts, if necessary
xylem—main water- and mineral-conducting tissue in trees and other plants; provides structural support and becomes wood after lignifying

ANSWERS TO WORKBOOK QUESTIONS AND SAMPLE TEST QUESTIONS

ANSWERS TO WORKBOOK QUESTIONS AND SAMPLE TEST QUESTIONS

Note: The Challenge Questions section in each chapter of the workbook consists of questions that require a higher level of knowledge and understanding to answer than the other workbook questions. At times, they draw on professional experience, other sources of information, or regionally specific information. Although the answers to these questions are not provided, the reader is encouraged to attempt these questions in order to develop a deeper understanding of the material and to be able to apply the knowledge to practical situations.

CHAPTER 1: TREE BIOLOGY

Workbook Questions

1. meristems
2. scales
3. apical dominance
4. leaves
5. photosynthesis, oxygen
6. chlorophyll
7. transpiration
8. stomata, guard cells
9. xylem, phloem
10. cambium
11. branch collar, branch bark ridge
12. bark
13. absorption
    conduction
    anchorage
    storage
14. osmosis
15. tropism, geotropism, phototropism
16. Compartmentalization Of Decay In Trees
17. excurrent, decurrent
18. mycorrhizae
19. respiration
20. deciduous, evergreen
21.

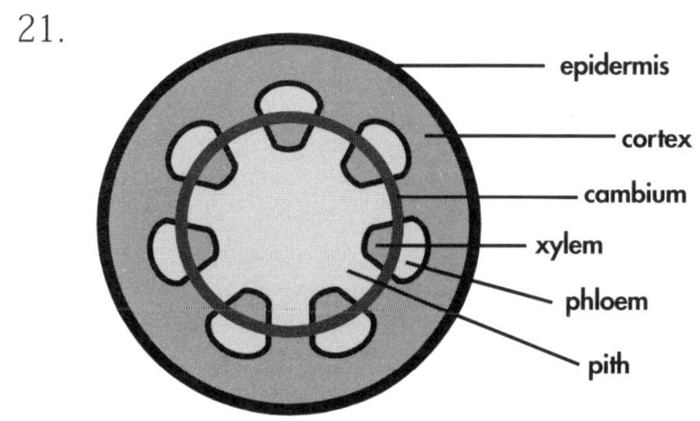

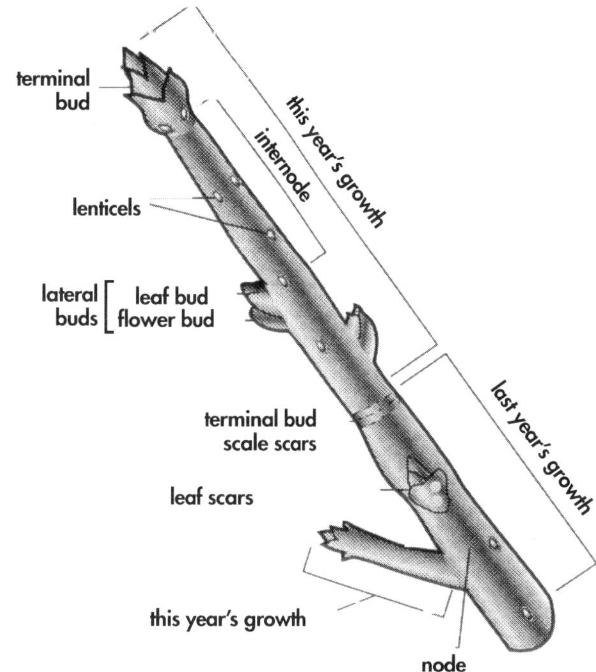

Matching

G. auxin
J. chlorophyll
E. cuticle
C. petiole
H. internode
F. lenticel
D. ray
B. absorbing roots
I. source
A. sink

Sample Test Questions

1. b
2. a
3. d
4. a
5. c

CHAPTER 2: TREE IDENTIFICATION

Workbook Questions

1. taxonomy
2. kingdom
   phylum
   class
   order
   family
   genus
   specific epithet
3. angiosperms, gymnosperms
4. monocotyledons (or monocots)
5. nomenclature
6. any five of the following:
   form or growth habit
   bark texture
   leaves
   flowers
   fruit
   seed
   buds
   leaf scars
   scent
7.

8. Palmately compound leaves: examples include buckeye, horsechestnut. Pinnately compound leaves: examples include ash, walnut, Kentucky coffeetree, ailanthus, honeylocust
9.

10.

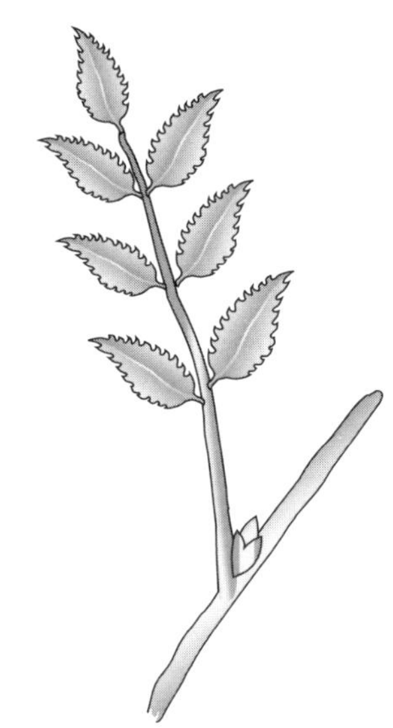

11. one
12. *Carpinus caroliniana* is one example.
13. genus, specific epithet
14. varieties, cultivars
15. cultivar

Sample Test Questions

1. a
2. a
3. b
4. c
5. c

CHAPTER 3: TREE/SOIL RELATIONS

Workbook Questions

1. O, A
2. compact, pore space
3. buffering capacity

4. true
5. one
6. soil structure
7. true
8. 2—silt, 1—clay, 3—sand
9. acidic, neutral, alkaline
10. true
11. 100
12. leaching
13. cations
14. alkaline
15. high
16. rhizosphere
17. false
18. gravitational, field capacity
19. false
20. true

Matching

F. sand
G. buffering capacity
H. field capacity
I. rhizosphere
J. macropores
A. mycorrhizae
E. CEC
C. clay
B. pH
K. micropores
D. gravitational water

Sample Test Questions

1. a
2. a
3. c
4. d
5. a

CHAPTER 4: WATER MANAGEMENT

Workbook Questions

1. true
2. late night, early morning
3. infiltration
4. water-holding capacity
5. compaction
6. true
7.

Advantages	Disadvantages
reduces water waste	must be moved outward as root system expands
may become clogged	reduces surface compaction

8. minimum irrigation
9. evapotranspiration
10. to moderate temperature extremes
    to reduce competition from weeds, grasses
    to keep mowers away
    to maintain soil moisture
    it is aesthetically pleasing
11. collar rot
12. antitranspirants
13. tensiometer
14. true
15. any three of the following:
    root suffocation/death
    soil organisms killed
    predisposition to other stress factors
    root collar rot
    tree prone to toppling

Sample Test Questions

1. a
2. b
3. c
4. d
5. d

CHAPTER 5: TREE NUTRITION AND FERTILIZATION

Workbook Questions

1. water
2. macronutrients
3. nitrogen
4. chlorosis
5. nitrogen, phosphorus, potassium
6. fertilizer analysis
7. 10
8. organic
9. slow-release
10. 2 to 4
11. water
12. true
13. aerating
14. fertilizer is placed below the absorbing roots
15. micronutrient
16. application wounds the tree
    wounds begin to coalesce if applied many times
17. fertilizer salts
18. leaching
19. element requirements
20. pH

Matching

D. micronutrients
F. 10-6-4
B. ureaformaldehyde
E. complete fertilizer
A. 2 to 4 pounds/1,000 square feet
C. chlorosis

Sample Test Questions

1. b
2. b
3. d
4. c
5. a

CHAPTER 6: TREE SELECTION

Workbook Questions

1. any species that grows higher than the wires would be a correct answer
2. hardiness
3. true
4. any three of the following:
   growing space
   light conditions
   soil conditions
   climate
   functional requirements
5. growth habits (or growth forms)
6. resistant
7. any three of the following:
   flowers
   attractive to birds
   fall color
   exfoliating bark
   growth habit
8. acclimation
9. false
10. any five of the following:
    plentiful, light-colored, healthy roots
    solid root mass
    good twig extension growth in previous years
    no major scars or injuries
    no insect or disease problems
    good branch structure

Sample Test Questions

1. d
2. b
3. d
4. a
5. c

CHAPTER 7: INSTALLATION AND ESTABLISHMENT

Workbook Questions

1. balled and burlapped
   bare root
   containerized
2. dormant
3. girdling
4. 2 to 3
5. burlap
6. deeper
7. shallow
8. gravel
9. true
10. true
11. early spring, fall
12. 10, 30 to 36 inches
13. false
14. true
15. root pruning
16. true
17. any three of the following:
    wounding of trunk or branches
    girdling if left in place too long
    less stable root development
    uneven trunk wood and taper development
18. upwind
19. true
20. moisture
21. water
22. slow-release
23. true
24. watering
25. true

Sample Test Questions

1. c
2. b
3. c
4. a
5. d

CHAPTER 8: PRUNING

Workbook Questions

1. growth
2. any five of the following:
   dead
   diseased
   hazardous
   crossing
   thinning for light or air penetration

weight reduction
size reduction
limbs are obstructing signs, views, clearance, etc.

3. leaf emergence, bloom
4. false
5.

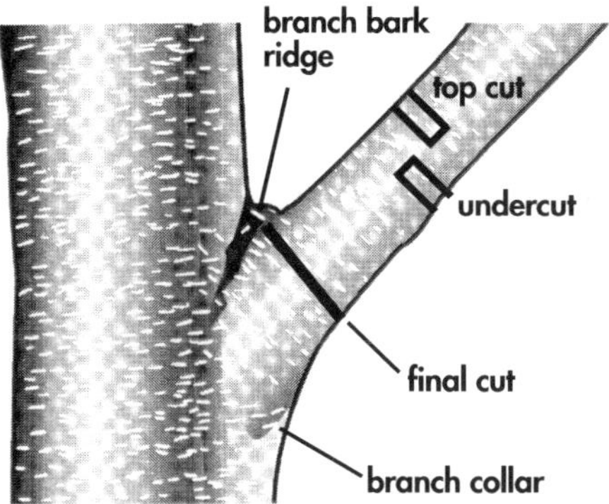

6. included bark
7. branch collar
8. codominant stems (or codominant branches)
9. true
10. true
11. subordinated
12. included bark
13. taper
14. 25
15. crown cleaning
16. true
17. thinning
18. lion tailing
19. uneven foliage distribution
    sunburned bark tissue
    weakened branch structure
20. reduction
21. excessive watersprout development
    weak branch attachment
    unnatural growth form
22. false
23. false
24. plant growth regulators
25. to reduce growth
    to restrict sucker growth

Sample Test Questions

1. d
2. b
3. d
4. c
5. a

CHAPTER 9: TREE SUPPORT AND PROTECTION SYSTEMS

Workbook Questions

1. support a weak crotch
   support for multi-stemmed tree
   weak-wooded tree
2. direct
   triangular
   box
3. pruned
4. two-thirds
5. taut
6. lag eye or lag hook
7. eye bolts, threaded rods with amon-eye nuts
8. true
9. 7-strand, common grade; extra-high-strength
10. thimble
11. dead-end grips
12. false
13. machine-threaded rods, lag-threaded rods
14. larger
15. round, oval
16. historic trees
    valuable trees
    large trees where people may seek refuge in a storm such as on a golf course
17. National Fire Protection Association, Lightning Protection Institute
18. air terminals
19. copper
20. 3
21. false
22. false
23. 10 feet
24. false
25. false

Sample Test Questions

1. c
2. c
3. c
4. d
5. a

CHAPTER 10: DIAGNOSIS AND PLANT DISORDERS

Workbook Questions

1. false
2. stressed
3. roots
4. true
5. abiotic
6. lawn mower/weed whacker
   vandalism
   construction
   rodents
   guy wires
7. egg laying
8. **Chewing** — **Piercing/Sucking**

Chewing	Piercing/Sucking
beetles	aphids
caterpillars	scales
ants	leafhoppers
leafminers	mealybugs
borers	whiteflies

9. vectors
10. true
11. nematodes
12. susceptible host
    pathogenic organism
    suitable environment
    proper timing
13. false
14. true
15. false
16. true
17. bacterium
18. allelopathy
19. sulfur dioxide
    fluoride
    ozone
    peroxyacetyl nitrates (PAN)
20. herbicide

Matching

D. witch's broom
B. vector
C. canker
A. gall
H. stunting
E. stress
F. pathogen
I. leaf spot
G. allelopathy

Sample Test Questions

1. a
2. a
3. b
4. b
5. a

CHAPTER 11: PLANT HEALTH CARE

Workbook Questions

1. false
2. vitality
3. vigor
4. growth
   maintenance
   storage
   defense
   reproduction
5. stress
6. cellulose, lignin
7. allelochemicals
8. enzymes
9. true
10. monitoring
11. appropriate response process
12. Integrated Pest Management
13. resistant
14. systemic
15. true
16. false
17. insect growth regulators
18. microbial extracts or microbial pesticides
19. true
20. predators, parasites

Sample Test Questions

1. c
2. b
3. a
4. c
5. c

CHAPTER 12: TREE ASSESSMENT AND RISK MANAGEMENT

Workbook Questions

1. potential for failure
   environment that may contribute to failure
   potential target
2. target
3. systematic

4. smaller
5. codominant
6. included bark
7. taper
8. reaction wood
9. branch union
10. decay
11. 30 to 35
12. decay
13. open wounds or cavities
    fruiting bodies: mushrooms, conks
    cracked or loosened bark
    certain insects
    birds, bees, other animals
14. excavation
15. mitigation
16. harm
17. removal of the tree or limb(s)
    pruning
    cabling
    bracing
18. negligence
19. act of God
20. higher standard

Sample Test Questions

1. d
2. b
3. d
4. a
5. c

CHAPTER 13: TREES AND CONSTRUCTION

Workbook Questions

1. root injury
   soil compaction
   injury to trunk or branches
   grade change
   excavation/severing root system
2. pore space
3. suffocation
   restriction of growth
4. lowered, raised
5. mulch
6. false
7. true
8. fences/barriers
9. bark tracing
10. aeration
11. false
12. terracing

Sample Test Questions

1. d
2. d
3. c
4. d
5. a

CHAPTER 14: SAFETY

Workbook Questions

1. shall, should
2. hard hat
   eye protection
   hearing protection, when applicable
   leg protection, when applicable
   sturdy work boots and appropriate clothing
3. true
4. true
5. ANSI Z133.1
6. false
7. false
8. true
9. true
10. "all clear"
11. job briefing
12. 10
13. emergency response
14. true
15. true
16. false
17. chain brake
18. false
19. true
20. true
21. false
22. open face
    traditional 45 degrees
    Humboldt
23. one-third
24. hinge
25. barber chair
26. side
27. true
28. shock
29. artificial respiration
30. poison ivy, poison oak, poison sumac

Matching

F. shall
G. approved
C. CPR
D. direct contact
B. should
H. indirect contact
E. ANSI Z133.1
A. chaps

Sample Test Questions

1. b
2. d
3. a
4. c
5. b

CHAPTER 15: CLIMBING AND WORKING IN TREES

Workbook Questions

1. manufacturers' guidelines
2. broken limbs
   electrical hazards
   dead limbs
   decay
   splits
3. climbing hitches
   tautline hitch
   Blake's hitch
   Prusik hitch
4. figure-8 knot, stopper knot
5. hitch, bend
6. throwing the line directly
   using a throwline
7. body thrusting, secured footlock
8. double crotching
9. safety lanyard (or work-positioning lanyard)
10. electrical hazard
11. rigging
12. double-braid
13. design factor
14. major axis
15. arborist blocks, rescue pulleys
16. false crotch
17. balanced
18. butt-hitching
19. drop cut
20. topping cut
21. (left to right) Blake's hitch, bowline, slip knot, figure-8

Matching

C. ANSI Z-133.1
H. scabbard
E. lowering device
B. Prusik loop
G. Blake's hitch
A. rope sling
F. access line
D. body thrust

Sample Test Questions

1. c
2. a
3. c
4. b
5. a

BIBLIOGRAPHY

BIBLIOGRAPHY

American National Standards Institute. *American National Standard for Tree Care Operations—Tree, Shrub and Other Woody Plant Maintenance—Standard Practices* (A300). ANSI, New York, NY.

American National Standards Institute. *American National Standard for Tree Care Operations—Tree, Shrub, and Other Woody Plant Maintenance—Standard Practices (Fertilization)* (A300, Part 2). ANSI, New York, NY.

American National Standards Institute. *American National Standard for Tree Care Operations—Tree, Shrub, and Other Woody Plant Maintenance—Standard Practices (Support Systems a. Cabling, Bracing, and Guying* (A300, Part 3). National Arborist Association, Manchester, NH.

American National Standards Institute. *American National Standard for Tree Care Operations—Pruning, Trimming, Repairing, Maintaining, and Removing Trees and Cutting Brush—Safety Requirements* (Z133.1). International Society of Arboriculture, Champaign, IL.

American National Standards Institute. *American National Standard for Nursery Stock* (Z60). American Association of Nurserymen, Washington, DC.

American Red Cross. 1992. *First Aid and Safety Handbook*. Little, Brown, Boston, MA.

Blair, D.F. 1999. *Arborist Equipment: A Guide to the Tools and Equipment of Tree Maintenance and Removal*. 2nd ed. International Society of Arboriculture, Champaign, IL.

Costello, L.R. 2000. *Training Young Trees for Structure and Form* (videocassette and booklet). University of California Cooperative Extension, Davis, CA.

Craul, P.J. 1999. *Urban Soils: Applications and Practices*. Wiley and Sons, New York, NY.

Dirr, M.A. 1998. *Manual of Woody Landscape Plants*. 5th ed. Stipes Publishing, Champaign, IL.

Esau, K. 1977. *Anatomy of Seed Plants*. 2nd ed. Wiley and Sons, New York, NY.

Foth, H.D. 1978. *Fundamentals of Soil Science*. 6th ed. Wiley and Sons, New York, NY.

Gilman, E.F. 1997. *An Illustrated Guide to Pruning Trees*. Delmar Publishers, Albany, NY.

Gilman, E.F. 1997. *Trees for Urban and Suburban Landscapes*. Delmar Publishers, Albany, NY.

Gilman, E.F. 1998. *Horticopia: Trees, Shrubs and Groundcovers* (CD-ROM). 2nd ed. Horticopia, Purcellville, VA

Hagen, B. 2000. Back to basics: Tree fertilization. *Arborist News* 9(6):34–41.

Harris, R.W., J.R. Clark, and N.P. Matheny. 1999. *Arboriculture: Integrated Management of Landscape Trees, Shrubs, and Vines*. 3rd ed. Prentice Hall, Upper Saddle River, NJ.

Hausenbuiller, R.L. 1978. *Soil Science: Principles and Practices*. 2nd ed. W.C. Brown, Dubuque, IA.

Himelick, E.B. 1991. *Tree and Shrub Transplanting Manual*. 2nd rev. International Society of Arboriculture, Champaign, IL.

International Society of Arboriculture. 1995. *Tree-Pruning Guidelines*. ISA, Champaign, IL.

International Society of Arboriculture. *ArborMaster Training Video Series I: Climbing Techniques and Equipment* (six videocassettes and workbooks). ISA, Champaign, IL.

International Society of Arboriculture. *ArborMaster Training Video Series II: Climbing Innovations* (two videocassettes and workbooks). ISA, Champaign, IL.

International Society of Arboriculture. *ArborMaster Training Video Series III: Chain Saw Safety, Maintenance, and Cutting Techniques* (six videocassettes and workbooks). ISA, Champaign, IL.

International Society of Arboriculture. *ArborMaster Training Video Series IV: Rigging* (seven videocassettes and workbook). ISA, Champaign, IL.

International Society of Arboriculture and National Arborist Association. 1999. *Basic Training for Tree Climbers* (five videocassettes and workbook). ISA, Champaign, IL, and NAA, Manchester, NH.

Jepson, J. 2000. *The Tree Climber's Companion*. 2nd ed. Beaver Tree Publishing, Longville, MN.

Johnson, G. 1999. Diagnosing abiotic disorders of landscape trees. *Arborist News* 8(4):31–37.

Johnson, W.T., and H.H. Lyon. 1991. *Insects That Feed on Trees and Shrubs*. 2nd ed. Comstock Publishing Associates, Cornell University Press, Ithaca, NY.

Kramer, P.J. 1969. *Plant and Soil Water Relationships: A Modern Synthesis*. McGraw-Hill, New York, NY.

Kramer, P.J., and T.T. Kozlowski 1979. *Physiology of Woody Plants*. Academic Press, New York, NY.

Lilly, S.J. 1994. *The Tree Worker's Manual*. Ohio Agricultural Educational Curriculum Materials Service, Columbus, OH.

Lilly, S. 1998. *Tree Climbers' Guide*. International Society of Arboriculture, Champaign, IL.

Lloyd, J., Editor. 1997. *Plant Health Care for Woody Ornamentals: A Professional's Guide to Preventing and Managing Environmental Stresses and Pests*. Cooperative Extension Service, University of Illinois at Urbana-Champaign, Urbana, IL, and International Society of Arboriculture, Champaign, IL.

Matheny, N.P., and J.R. Clark. 1994. *A Photographic Guide to the Evaluation of Hazard Trees in Urban Areas*. 2nd ed. International Society of Arboriculture, Champaign, IL

Matheny, N.P., and J.R. Clark. 1998. *Trees and Development*. International Society of Arboriculture, Champaign, IL.

Merullo, V.D., and M.J. Valentine. 1992. *Arboriculture and the Law*. International Society of Arboriculture, Champaign, IL.

Miller, F. 2000. Want to be a better plant diagnostician? *Arborist News* 9(4):33–39.

Miller, R.H. 1999. Soil properties, part 1. *Arborist News* 8(5):56–61.

Miller, R.H. 1999. Soil properties, part 2. *Arborist News* 8(6):14–18.

National Arborist Association and International Society of Arboriculture. 2000. *Basic Training for Ground Operations in Tree Care* (five videocassettes and workbook). NAA, Manchester, NH, and ISA, Champaign, IL.

Partyka, R.E., J.W. Rimelspach, B.G. Joyner, and S.A. Carver. 1980. *Woody Ornamentals: Plants and Problems*. ChemLawn, Columbus, OH.

Pritchett, W.L. 1979. *Properties and Management of Forest Soils*. Wiley and Sons, New York, NY.

Rost, T.L., M.G. Barbour, R.M. Thornton, T.E. Weir, and C.R. Stocking. 1984. 2nd ed. *Botany: A Brief Introduction to Plant Biology*. Wiley and Sons, New York, NY.

Salisbury F.B., and C.W. Ross 1978. *Plant Physiology*. 2nd ed. Wadsworth Publishing, Belmont, CA.

Shigo, A.L. 1986. *A New Tree Biology*. Shigo and Trees, Associates, Durham, NH.

Shigo, A.L. 1986. *A New Tree Biology Dictionary*. Shigo and Trees, Associates, Durham, NH.

Shigo, A.L. 1989. *Tree Pruning: A Worldwide Photo Guide*. Shigo and Trees, Associates, Durham, NH.

Sinclair, W.A., H.H. Lyon, and W.T. Johnson 1987. *Diseases of Trees and Shrubs*. Comstock Publishing Associates, Cornell University Press, Ithaca, NY.

Sunset Editors. 1995. *Western Garden Guide*. 6th ed. Sunset Publishing, Menlo Park, CA.

USDA Forest Service. 1996. *Urban Forestry Laboratory Exercises: For Elementary, Middle, and High School Students*.

Watson, G.W., and E.B. Himelick. 1997. *Principles and Practice of Planting Trees and Shrubs*. International Society of Arboriculture, Champaign, IL.

Watson, G.W. No date. *Root Injury and Tree Health* (videocassette and booklet). Illinois Arborist Association, Antioch, IL.

Whitcomb, C.E. 1991. *Establishment and Maintenance of Landscape Plants*. Lacebark, Stillwater, OK.

INDEX

INDEX

Page numbers in **boldface** refer to figures. Page numbers followed by (t) refer to tables.

C

(continued)

J

K

L

T